Location		
Main Class No. 5 - 905		X
542	H	X
Added Class Nos.	R	
WITHDRAWN	P	X
	Q	
Accession No.	T	X
18460	W	
14/12/79		
	X	✓
RECKITT & COLMAN Library and Information Service		

D1793005

Laboratory Organization and Management

Laboratory Organization and Management

Fred Grover, FIST
Safety Adviser, Medical Research Council

Peter Wallace, FIST
Laboratory Superintendent, Nuffield Laboratories
of Comparative Medicine, Institute of Zoology,
The Zoological Society of London; Safety Adviser,
The Zoological Society of London

Butterworths
London Boston
Sydney Wellington Durban Toronto

United Kingdom	Butterworth & Co (Publishers) Ltd	
London	88 Kingsway, WC2B 6AB	
Australia	Butterworths Pty Ltd	
Sydney	586 Pacific Highway, Chatswood, NSW 2067	
	Also at Melbourne, Brisbane, Adelaide and Perth	
Canada	Butterworth & Co (Canada) Ltd	
Toronto	2265 Midland Avenue, Scarborough, Ontario M1P 4S1	
New Zealand	Butterworths of New Zealand Ltd	
Wellington	T & W Young Building, 77-85 Customhouse Quay, 1, CPO Box 472	
South Africa	Butterworth & Co (South Africa) (Pty) Ltd	
Durban	152-154 Gale Street	
USA	Butterworth (Publishers) Inc	
Boston	10 Tower Office Park, Woburn, Massachusetts 01801	

All rights reserved. No part of this publication may be reproduced or transmitted in any form or by any means, including photocopying and recording, without the written permission of the copyright holder, application for which should be addressed to the Publishers. Such written permission must also be obtained before any part of this book is stored in a retrieval system of any nature.

This book is sold subject to the Standard Conditions of Sale of Net Books and may not be re-sold in the UK below the net price given by the Publishers in their current price list.

First published 1979
© Butterworth & Co (Publishers) Ltd, 1979
ISBN 0 408 70793 3

British Library Cataloguing in Publication Data

Grover, Fred
    Laboratory organisation and management.
    1. Laboratories — Administration
    I. Title II. Wallace, Peter
    658'.91'50724   Q183.A1   79—40289

ISBN 0-408-70793—3

Typeset by Scribe Design, Medway, Kent
Printed and bound at the University Press, Cambridge

Preface

In recent years much has been written on the theory of management, but many will consider that some of the academic and often elegant theories that have recently been propounded have little relevance to the purely practical world of the laboratory. Indeed, a number of those technicians who have moved into laboratory management following a successful earlier career at the laboratory bench will argue that management training, as it is generally known and taught, is neither relevant nor appropriate to the needs of the modern laboratory.

There can be no doubt that good management increases the efficiency of any organization, be it commercial or academic. In the competitive society of today, it is those organizations which lack good management which are most likely to become the lame ducks and suffer the inevitable failure which results from this weakness.

The purpose of this book is to survey the practical problems which the laboratory manager is likely to encounter in his daily task and to offer some guidelines to their solution. Management is about getting things done, and to this end the manager's job is to ensure that the staff have the facilities and materials necessary for them to do their work, that budgets are not exceeded, that hazards are reduced to the absolute minimum, that legislation affecting the work is not infringed and that the whole organization runs as smoothly as possible.

Major changes have taken place in laboratories over the years; many laboratories are now enormous buildings staffed by highly trained scientists, technologists and other supporting staff, with budgets comparable to those relating to a small town. The role of the technician has grown with these changes. No longer is he the fetcher and carrier, bottle filler and bench polisher of the past. Laboratory work has become so complex and equipment so sophisticated that new attitudes as well as new skills are required of him.

Management can be divided into two overlapping sections. One of these consists in devising and operating systems of work, whilst the other is concerned with getting the best out of people. Both are dedicated to helping the work along and not suffocating it with bureaucracy (there are bureaucrats in plenty to do this for us).

Not every technician or scientist makes a good laboratory manager and to many the very thought of management as a career is repugnant. Unhappily it is a fact that as a supervisor climbs the career ladder he has an increasing burden of administrative work thrust upon him. Many good technicians, at the peak of their working lives, find they must forsake the bench for the desk, but nevertheless, successful laboratory management does require a solid background of laboratory experience. Managers trained in other spheres have occasionally been appointed as laboratory managers and whilst some have achieved greatness, the results have in many instances been disappointing.

In the long term, it may prove possible to modify technical training very substantially, hence enabling those who so wish to stay at the bench; those whose interest lies in management, however, will be encouraged to follow this interesting and rewarding career.

The task of laboratory management varies considerably from one organization to another and it is therefore not possible for a book such as this to provide hard-and-fast rules concerning what should be done in a given set of circumstances. Rather, it is intended to act as a practical guide to those setting out in laboratory management and perhaps to give further food for thought to those already involved.

F.G.
P.W.

Acknowledgements

Before our paths converged at the University College Hospital Medical School we were each fortunate to have received a very broad training in industrial chemistry with companies who were the leaders in their particular fields. Subsequently, when we changed direction and came together at UCHMS we were even more fortunate to fall under the influence of, and be guided by, one of the most effective and highly regarded chief technicians in the profession. That person, Mr A.W. Hemmings MBE, who recently retired from the National Institute for Medical Research, taught us the very basics of our trade and by his inspiration and example as a manager encouraged us to study and to develop our careers. Many of his ideas and philosophies have been incorporated in this book and we will always be grateful for his kindly advice and for his friendship throughout our working lives.

Paul Diamond of the Biochemistry Department at The Royal College of Surgeons of England made the original suggestion that we should write this book and we are grateful for his encouragement.

Our publishers' commissioning editor has been a tower of strength throughout the production period and we are most grateful for her kindness and understanding. Never once did she complain over the inevitable delays in producing the final manuscript.

John Massey, an inspired teacher of one of us (FG) and long time friend of both, thoroughly and painstakingly read the first draft and made many suggestions that we were happy to incorporate.

To our wives who sustained us in our prolonged efforts to produce the manuscript and without whose encouragement we would never have completed the work, we offer our humble thanks.

We are especially grateful to Tony Hoare of A.R. Hoare & Co. Ltd, and to his publicity agent Gerald Bishop of Quantum Technical Communications Ltd, who provided the original photographs for this book. We also gratefully acknowledge the co-operation of their clients whose premises and equipment were photographed.

Mr D. Busby of the National Institute for Medical Research kindly read the proofs and his keen eye has eliminated many of the mistakes; for any that remain we accept the entire responsibility.

Last, and by no means least, we offer our thanks and apologies to our many friends, colleagues and acquaintances who have provided information and help which has been freely adapted for our purpose.

F.G.
P.W.

Contents

1 *Laboratory Planning and Layout*   1
Preliminary considerations; space requirements; laboratory layout; provision of services; floors; windows; doors; benches; cupboards and drawer units; mechanical services — heating and ventilating, fume cupboards, lighting, electrical supplies, gas supplies, water supplies, piped gases; safety in design; decoration; allocation of floors in multi-storey buildings

2 *Selection and Management of Staff*   33
Job description; the advertisement; application forms — references; interviewing and selection — final selection; contracts and conditions of service; induction; training and further education — motivation in technical education, recent developments in technical training and education; laboratory discipline; termination of employment

3 *Purchasing and Financial Control*   62
Sources of income; purchasing — capital equipment, consumable materials, ordering procedure, special purchasing considerations, cost cutting

4 *Management of Stores*   76
Stores policy; stores design and planning — storage of chemicals, hazardous materials, storage of apparatus; documentation

5 *Laboratory Administration* 95
Technical information — filing systems, indexing systems, Dewey Decimal Classification; laboratory records and record-keeping; office facilities and equipment; decision-making — seeking advice, staff meetings, advisory committee meetings, formal committee meetings

6 *Service Departments and Special-purpose Rooms* 111
Glassware washing and sterilizing facilities; radioisotope laboratories; photographic units; cold-rooms; hot-rooms; animal houses; reprographic units; laboratory workshops; audio-visual aids — audio aids, visual aids; glassblowing shops

7 *Health and Safety* 146
The basic approach — The Health and Safety at Work etc. Act 1974; organization of laboratory safety — line management, safety officers, safety committees, codes of practice, general attitude to laboratory safety, accident books and records, notifiable accidents; the hazards — fire, fire prevention, fire-fighting equipment, fire drills, fire escapes, fire prevention advice; electrical and electronic equipment; radiation and the use of radioactive substances; cylinders of compressed gas; centrifuges; cryogenic substances; physical injuries; chemicals; occupational hygiene; dermatitis and skin reactions; toxic substances and threshold limit values; carcinogens; bacteria, viruses and other biohazards

8 *Maintenance of Laboratory Premises and Equipment* 177
Planned maintenance — inspection of premises and equipment

9 *Automation in the Laboratory* 186

10 *Management Techniques and Functions* 192
Forecasting; planning; organizing; motivating; co-ordinating; controlling; communicating

Appendix 1 Certificate in Medical Laboratory Management 211

Appendix 2 Diploma in Laboratory Management 217

Appendix 3 Technician Education Council Higher Awards in Laboratory Science and Administration 223

Appendix 4 List of Papers Available from the Laboratories Investigation Unit 232

Appendix 5 Safety Legislation in EEC Member States 234

Index 237

1
Laboratory Planning and Layout

Preliminary considerations

During the course of his career the laboratory manager is certain to become involved in laboratory design, either with the preparation of plans for a new building or extension, or in the conversion or updating of existing laboratories. In order to co-operate successfully with architects and contractors, it is essential for him to have a working knowledge of the basic principles of design and layout.

When a new building is being designed, an architect is appointed to supervise all aspects of planning and building. It is his function to ensure that the finished building meets the requirements of the client and complies with local and national legislation. In order to achieve this, it is essential to discuss these requirements in great detail with the client and exchange views on how the various problems can best be overcome. Any plan must, of necessity, be something of a compromise between what the client wants, what is possible in terms of construction, the limitations imposed by the site and the finance available. It is impossible to overemphasize the importance of good planning in the early stages, as modifications at a later date are often impossible and always expensive.

It is usual to appoint a small planning committee consisting of representatives of the architects, the builders and the client (including the laboratory manager). A brief is prepared setting out in detail the nature of the work to be done in the building, the number of employees expected to work in each department, requirements for special rooms, dimensions and weight of major equipment, the number of floors required, mechanical services, access from the road, national and local legislation which may apply, and so on.

Visits to other laboratories engaged in similar work are always useful if only to avoid errors that may have been made in the past. One can also consult the staff in these buildings who have had the experience of working in them and discovering the problems not obvious at first sight. It is frequently the easily overlooked details which become sources of frustration and inefficiency in a building which in all other respects is well designed.

A laboratory building is costly and is intended to last for many years, during which time the nature of the work is almost certain to change, and hence some degree of flexibility is desirable in order to make updating possible. This may be achieved by the use of demountable partitions or lightweight non-load-bearing walls in some areas so that, for example, two small laboratories can fairly easily be made into one large one. Most construction today is in modular form, i.e. dimensions are standardized throughout to include room sizes, lengths of bench units, etc., such dimensions being usually based on a 500 mm unit. Thus a room may be square, e.g. 4.5 m by 4.5 m or rectangular, e.g. 7 m by 3.5 m and bench units of 1.5 m length would fit into either room (*Figure 1.1*).

There are advantages to both the square and rectangular layouts. The former makes better use of natural daylight and tends to be more flexible in terms of furniture arrangement, whilst the latter allows longer runs of bench surface if doors and windows are on the shorter walls. Some rooms do not require daylight and these can usually be sited in the centre of the building.

It is important to remember that rooms other than laboratories will be required (e.g. stores, offices, library, cold storage facilities, etc.) and that a considerable floor area is occupied by corridors, staircases, landings, lifts, entrance hall, etc., which may well account for as much as 30% of the total space. From a safety aspect, these areas must be adequate to allow rapid escape in the event of an emergency and corridors must be wide enough to allow easy passage of trolleys.

Staff also have to circulate and a knowledge of the main

functions of each department will assist in deciding how they should be accommodated in relation to one another. Some activities will need to be sited close to each other, whilst other may benefit by being separated (e.g. to reduce the hazard of cross-infection).

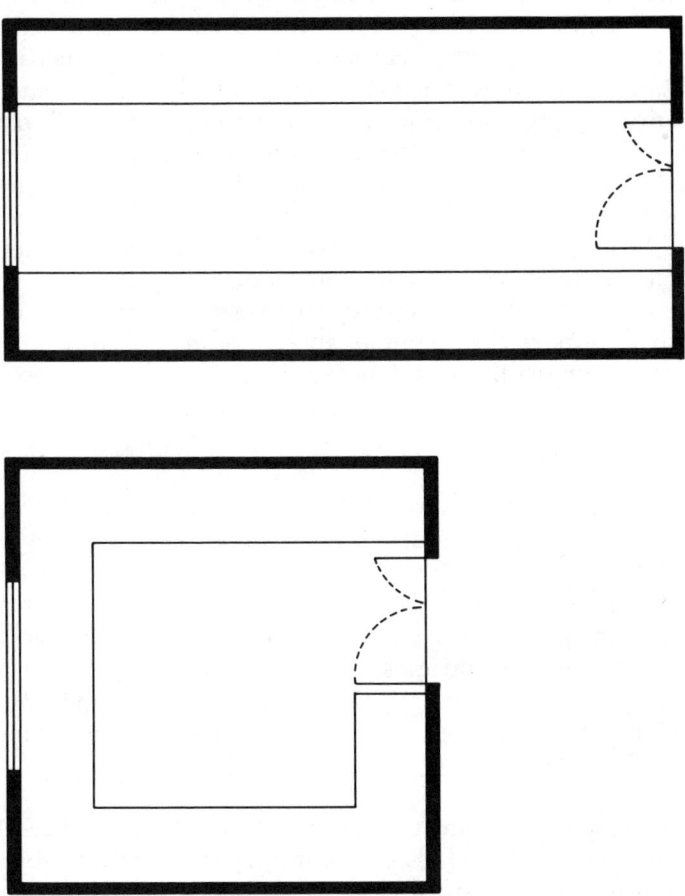

Figure 1.1. *Square and rectangular modular laboratories — each has one window and a 'door and a half'*

The architect's attention should be drawn to any unusual hazard that may not be apparent to him, such as high-level radioactive work or the housing of animals under controlled conditions, and a decision can be made as to whether special areas are set aside for this work or if purpose-designed cabinets or isolators should be installed. The problems of goods delivery, waste disposal, bulk storage, temperature-controlled rooms, glassware washing, etc., must also be discussed in detail as they may be of considerable importance in the day-to-day work. Armed with this information, the architect will be in a position to prepare a provisional specification and plan which can then be discussed and modified as necessary. He may also appoint specialist consultants to handle certain aspects of the work, such as interior fittings and mechanical services. When all parties are satisfied with the design, authorization is given for the work to be started. Unfortunately, inflation invariably necessitates some modifications being made as building proceeds and it may be possible to limit them to minor deletions which can be made good at a later date.

To enable the laboratory manager to make a useful contribution at the planning stage, the main features of the laboratory must be considered in detail.

Space requirements

These requirements vary with the type of work being done, but the following suggestions form a useful guide:

Research laboratories	20—25 m^2 per worker
Routine diagnostic or analytical laboratories	15—20 m^2 per worker
School teaching laboratories	2½—3 m bench per student
University teaching laboratories	2—6 m bench per student

Storage space outside the laboratories will also be required to the extent of 8—10% of laboratory floor area. In teaching departments, laboratories can often be used as lecture rooms also, but in other establishments this is seldom practicable and an area may have to be set aside for discussions, seminars and similar activities.

Laboratory layout

The basic plan for laboratories has evolved over many years and since the 1920s the development has been based on the concept of rooms arranged on each side of a central corridor[1]. This system has much to commend it, since it is very economical in space, improves communication from one room to another and simplifies the provision of laboratory services. It is based on a structural module, the size and shape of which is designed to meet the requirements of the individual worker. Benches are at right angles to the window wall and provide uninterrupted runs. Daylight is utilized to the maximum and rooms can therefore be deeper than would otherwise be acceptable. The modules can be subdivided by lightweight partitions to provide offices, store rooms, instrument rooms, etc. (*Figure 1.2*).

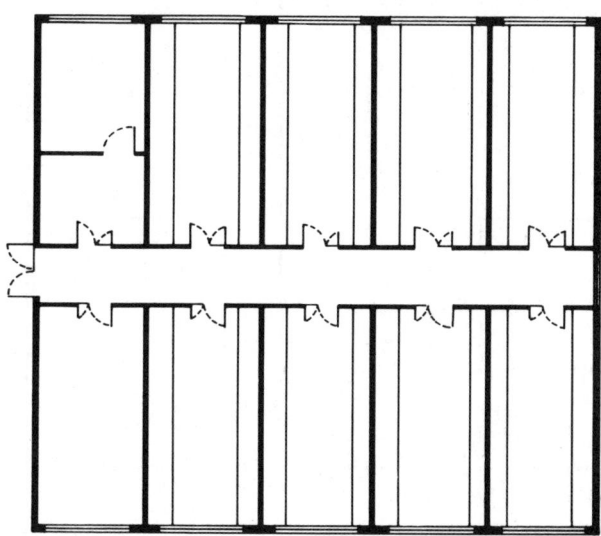

Figure 1.2. Laboratories arranged on either side of a central corridor. The dividing walls may be demountable

If the site is of suitable dimensions, a two-corridor system of similar modules is possible (*Figure 1.3*).

The area between the corridors is used for rooms which do not require daylight, such as cold stores, photographic dark rooms, etc. Staircases and lifts, where required, are constructed at the ends of the building. The dividers between laboratories are non-load-bearing partitions which can be removed to provide double- or treble-sized rooms if necessary. Intercommunicating doors between laboratories

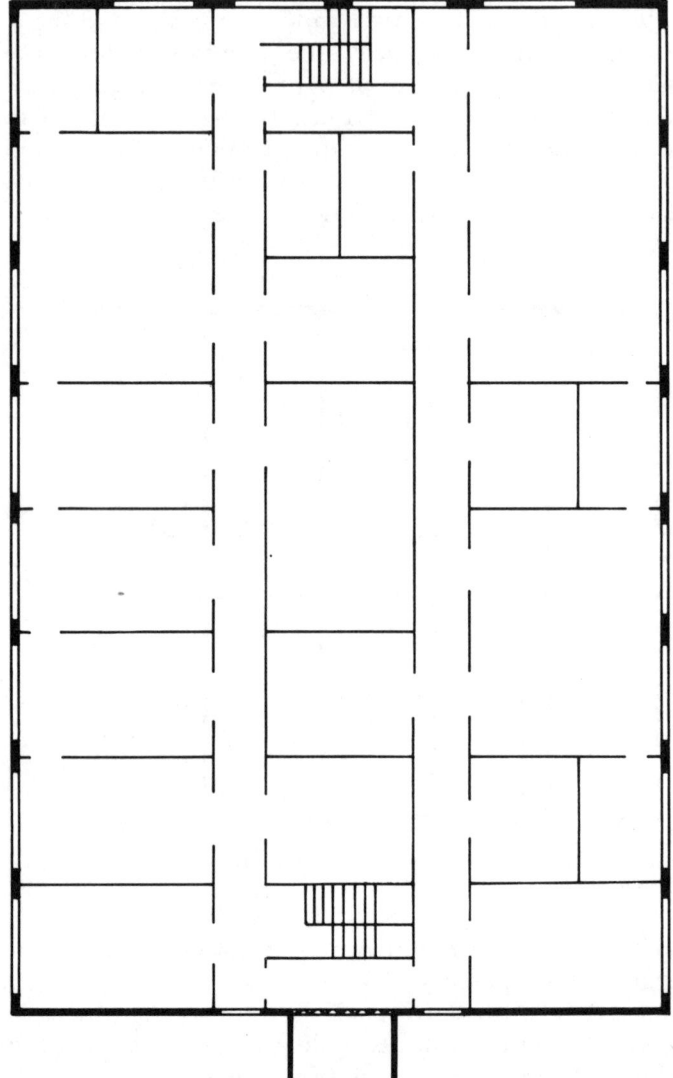

Figure 1.3. Laboratories arranged on a two-corridor system. The central core is used for storage, toilets, photographic dark rooms, etc.

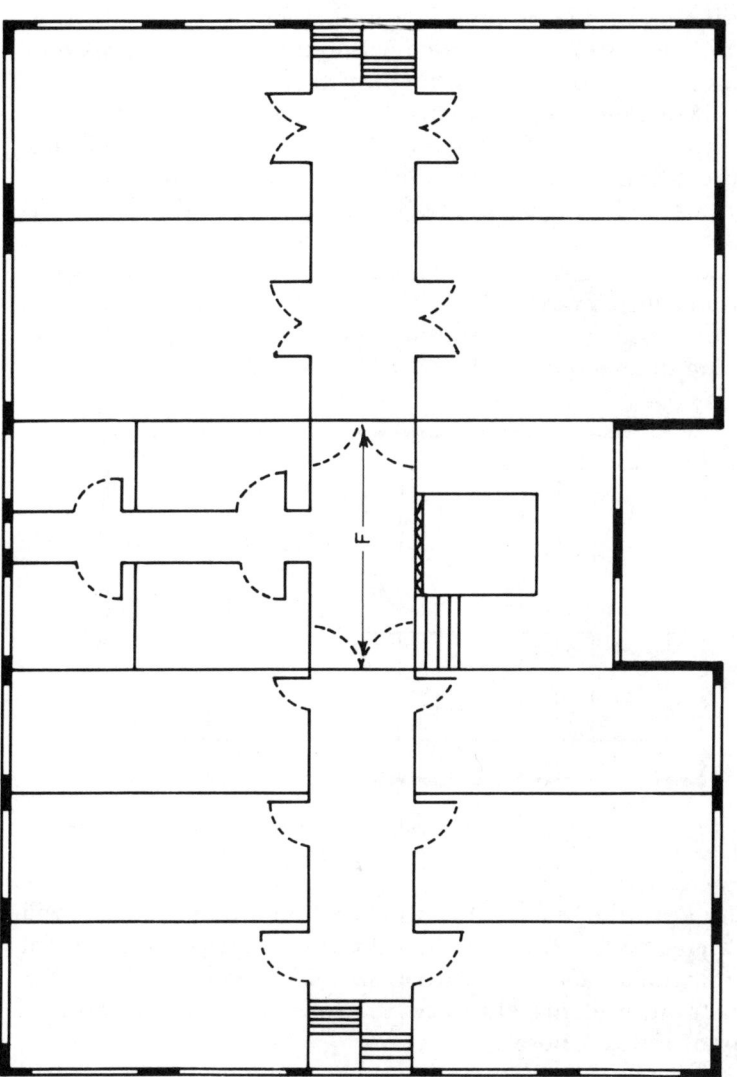

Figure 1.4. An arrangement suitable for a teaching establishment, in which laboratories and lecture rooms are on either side of a central stair well and lift shaft (F, fire doors)

are not only useful in everyday situations but also provide a second exit in the event of a fire. The provision of two staircases, one at each end of the floor, also ensures a means of exit should one be unusable during an emergency (this is usually a statutory requirement). Should this not be possible, an external fire escape must be provided.

In teaching departments, the lecture rooms and laboratories are usually separated by a central staircase. This has the advantage of noise isolation, and in an emergency the laboratory side can be shut off rapidly by fire doors (*Figure 1.4*).

In large laboratories, the bench units may be used to divide the room into working bays giving each worker his own area but retaining good communication between staff by not isolating them from one another (*Figure 1.5*).

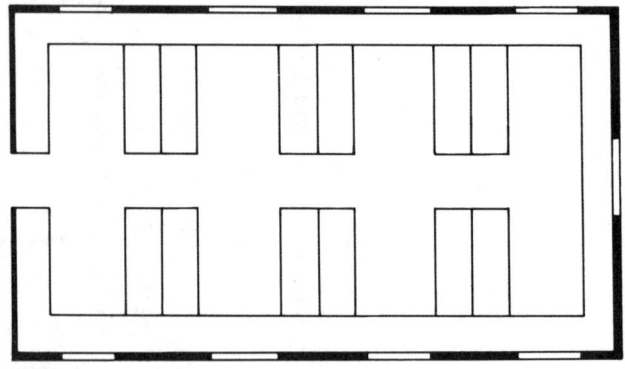

Figure 1.5. A large laboratory divided into working bays by peninsular bench units

The individual laboratories in teaching establishments will in general be much larger than those in research or industrial organizations, and the actual size is calculated from the number of students expected to work in them and from a study of the syllabuses.

Experience has shown that once a building is occupied there is a tendency for the number of students to increase over the years and the resulting overcrowding is both frustrating and hazardous. The corresponding increase in staff will also aggravate the problem. Where possible, allowances should be made for this eventuality at the planning stage, or alternatively, provision must be made for extensions to be added in the future. Such extensions could take the form of an additional floor or a lengthening of the building. The latter is preferable as it can be done with the minimum of disruption.

Provision of services

The methods used to carry services from the incoming supply to service points at the benches vary enormously and a great deal of thought has been put into this problem over the years by architects, mechanical service engineers and laboratory users. Ideally the runs should be kept as short as possible, be readily adapted for alterations and extensions, be easily accessible for maintenance, be protected from accidental damage and tampering, should not be in the way of benching, apparatus, etc., and be kept out of sight. It is doubtful if any system has yet been devised which fulfils all these criteria but some at least approach the ideal.

Services should rise upward from the intake level and be distributed horizontally at each floor level in order to feed benches in each laboratory. Current practice is to install service ducts for rising mains and carry branch pipes in the voids between floors or behind bench units. Ducts and voids are provided with access points for servicing. The mass of pipework and cabling in laboratory buildings is so great that space required for them must be designed into the building, and can represent a significant percentage of the total volume.

Corridor walls can be utilized for access doors to rising mains at the points where they are connected to horizontal sub-mains feeding the laboratories. The same space may house ducting for fume cupboards or general ventilation.

Service pipes to the bench outlets may be below bench level, above bench level in a service spine or dropped from the ceiling void; waste pipes must, of course, be below bench level.

The service spine is a very satisfactory arrangement but requires the bench to be set away from the wall about 15 cm, with the result that the incorporation of this feature produces, on each side of a room, a significant loss of width between benches. The main advantage to dropping services from the ceiling void is that partitions between rooms may be removed without disturbing the pipework. It does, however, require larger ceiling voids to provide access for maintenance together with duct covers in the floors above, and these are difficult to keep watertight.

A building consisting of more than one floor will require at least one lift, which should have a capacity of at least one tonne. It is common practice to install one lift for passengers and another for goods, although in many cases a single lift will suffice provided it is of adequate capacity.

Floors

The architect will decide on the type of floor construction, having due regard to its load-bearing capacity. It is important to remember that vibrating equipment can represent a load of two or more times its static weight. It may be desirable to have some sections of floor isolated from their surroundings to prevent vibration from one piece of equipment affecting other apparatus. For example, an electron microscope could not be operated satisfactorily if it were not isolated from a nearby heavy-duty freezer compressor.

Floor covering material is important from a number of aspects, e.g. safety, comfort, ease of maintenance, resistance to chemical and solvent attack, life expectancy, cost and appearance. There is today a wide choice of materials available. Traditionally, linoleum has been most commonly used and still has much to recommend it. It is available in wide rolls (which minimizes the number of joints) and fulfils all the requirements mentioned above. Vinyl coverings are also popular but tend to be slippery when wet and are attacked by some solvents.

Areas which are frequently wet, such as animal rooms, are usually covered with asphalt, but this is easily dented and may give rise to pockets where infected materials may lodge. Quarry tiles are useful in these areas provided they are well laid and grouted.

Hardwood is a pleasing finish for staircases and corridors but unsatisfactory in laboratories owing to the large number of joints and the high cost of maintenance. Offices, libraries and similar areas may be covered with synthetic fibre carpet which is quiet, comfortable, hard-wearing and inexpensive.

Workshops and other areas where mains voltage equipment is being used under difficult conditions must be provided with a floor covering which is a good electrical insulator — linoleum is generally considered to be satisfactory in these areas (see also Chapters 6 and 7).

Windows

Most laboratories require large window areas to enable natural lighting to be utilized as much as possible, to provide at least part of the ventilation and, in some instances, to provide a means of escape in emergencies. They should be easily opened by the smallest person likely to occupy the room, be readily cleaned with minimum disturbance to the occupants, be fitted with reliable closures which prevent access by unauthorized persons, and incorporate frames which are resistant to the corrosive atmosphere present in some laboratories. The siting of windows must be chosen carefully as they occupy wall space which could be used for other purposes. Two smaller windows may be less wasteful of space than one large one. Double glazing, although costly, may be an advantage if the normal ambient temperature is particularly low for long periods of the year. It also assists in the reduction of noise from traffic, etc.

Doors

The usual practice is to provide laboratories with a 'door and a half', i.e. two leaves, one larger than the other. The large leaf serves as the normal entrance and the small one is also opened when large items need to be taken through the doorway. Suitable widths are 90 cm and 45 cm.

Laboratory doors should be fitted with glass panels to enable one to see if it is safe to open the door and to enable a visual check to be made on the room without opening the door. It may be desirable to arrange for all laboratory doors to be lockable and local regulations may require certain doors to be locked when not in use, e.g. flammable solvent stores and dangerous chemical stores. All locks should operate on a 'master key' system.

Benches

The choice of bench units will depend on the degree of flexibility required. Fixed benching is acceptable where the techniques used are unlikely to alter, but if the type of work is likely to change fairly rapidly some degree of flexibility is desirable, bearing in mind the increased cost. Modular units are extremely flexible but do not provide an uninterrupted working surface. Floor-standing equipment is readily accommodated by removing one or more units[2].

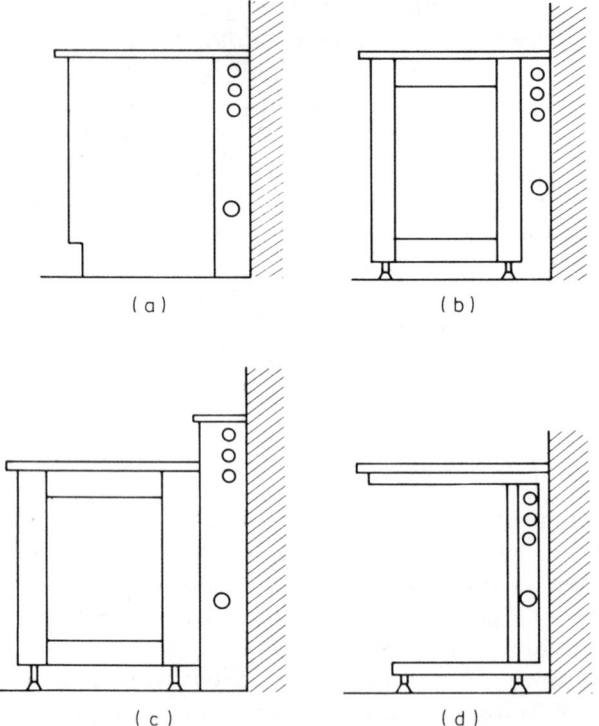

Figure 1.6. Services fitted behind under-bench units: (a) services virtually inaccessible; (b) removable bench unit giving improved access; (c) services run in a spine installed behind a free-standing bench unit; (d) cantilever bench construction showing cupboard unit removed for access

Figure 1.7. Under-bench services. Plumbing is incorporated into a service spine prior to bench installation, permitting easy access for maintenance (courtesy A.R. Hoare & Co. Ltd)

Whichever units are chosen, rigid construction is essential. Timber frames and worktops are the most commonly used, but metal units are becoming increasingly popular as corrosion-resistant finishes of improved quality become available. Floor-standing frames are usually cheaper but brackets cantilevered from the walls have the advantages of giving an unobstructed floor space and better access to service pipes (*Figures 1.6* and *1.7*). Bench units are available in heights suitable for standing or sitting work, e.g. 75 and 90 cm.

A wide variety of benchtop material is in used, hardwood being the most popular. Regrettably, top grade teak is becoming scarce and very expensive and whilst some of the alternative hardwoods are satisfactory, many are not. Several factory-made timbers, such as blockboard, are well suited as benchtop materials provided they are covered with an impervious plastic laminate fixed with a good-quality adhesive. These materials wear well, are easy to keep clean and are resistant to most chemicals. They can, however, be damaged by heat or impact but are not difficult to replace. In laboratories where infected materials are handled, the smooth finish of the laminate is readily disinfected.

Linoleum work surfaces are excellent in areas such as instrument rooms and are easily replaced if damaged. Sheet lead may be used where corrosive acids are handled in quantity, but the cost is so great that it is doubtful if such expense can be justified. Quarry tiles are suitable for this type of work but must be fixed and grouted with acid-resistant products.

Cupboards and drawer units

The bulk of laboratory storage is in under-bench cupboards, although wall-mounted units are used to house small, light items. As with benches, these may be constructed of timber or metal. Under-bench units are usually floor-standing and arranged so that they can be pulled out for cleaning and for access to services. The units may also be suspended from the bench top, in which case they usually slide sideways to gain access to services. Sliding doors are most suitable for wall-mounted cupboards but hinged doors are more satisfactory for under-bench units. Drawer units are useful for small items in frequent use at the bench. Good-quality workmanship is vital, because badly made drawers tend to stick and their contents may be damaged when trying to force them open.

Fire-resistant cabinets for the storage of flammable solvents should be provided in those laboratories where such materials are in use. Units are available with built-in automatic fire extinguishers. These should be large enough to hold one bottle of each solvent required but not so large as to encourage the keeping of large volumes which should be stored in a proper solvent store (see Chapters 4 and 7).

Other furniture, such as desks, chairs, filing cabinets and lockers, should be provided where necessary. Desks and filing cabinets should be lockable; the latter should be of fire-resistant construction. In some laboratories, a safe may be required for the storage of confidential documents, or for materials which are valuable or dangerous.

Mechanical services

The main difference between a laboratory building and any other building is in the type of mechanical services required, and the correct provision of these services is of paramount importance[3].

HEATING AND VENTILATING

It is generally accepted that 20°C is a suitable temperature for most laboratories. A higher temperature tends to be oppressive and a lower temperature is uncomfortable and reduces manual dexterity. Heating is usually provided by hot water radiators fed from a central boiler or by ducted warm air. The latter system may be unsuitable in some laboratories as it introduces a possible hazard where infected or other dangerous material is involved. In instances where the ambient temperatures are likely to be above 20°C for long periods of the year, or where a large amount of heat is produced in the building, full air conditioning will be necessary.

Figure 1.8. Ventilation plant showing fans and ducting for air extraction from a number of laboratories. The extracted air is passed into a common manifold and used to dilute the extract from fume cupboards and is finally exhausted via a 10 m high vertical chimney at roof level (courtesy A.R. Hoare & Co. Ltd)

In most laboratories, some form of mechanically assisted ventilation is desirable to maintain a healthy environment (*Figure 1.8*). It may not be necessary to use this continually but it should be installed in areas where noxious or dangerous fumes are likely to be generated. Many volatile materials are dangerous if inhaled for long periods (see Chapter 7) and good ventilation helps to keep these to a minimum concentration. Ventilation is measured in air changes per hour, an average figure for laboratories being 6. If large amounts of noxious fumes or gases are produced, it is usually better to provide efficient fume cupboards rather than to overcome the problem by increasing whole-room ventilation. A ducted warm-air supply is provided by one or more fan units working in conjunction with a balanced air extract system. The latter will of course remove heat from the building and this heat loss must be allowed for in the calculations for the heating requirements.

Some rooms may require their air supply to be filtered and this is normally done by installing disposable filter elements in the supply ducts. These should be capable of removing particles of diameters larger than 5μm and will need to be replaced regularly. Electrostatic dust removers are efficient but are expensive to install and require considerable maintenance.

FUME CUPBOARDS

These cupboards will be required in some laboratories, particularly those involved in chemistry or radioactive work. Until a few years ago, most fume cupboards were more decorative than functional as their performance was so inadequate; current models available from specialist manufacturers are, however, extremely efficient and are correctly matched to the ducting and extract fans.

The rate of extract required will vary with the type of work for which the fume cupboards are used. This rate is measured in terms of air velocity over the working surface. For general work, a minimum of 0.4 m/s is recommended and for work with highly toxic or radioactive material, a minimum velocity of 0.5 m/s is required. These velocities are measured with the sash open to a height of 600 mm. Too high a velocity can cause light powders to be drawn into the extract system.

It is not generally appreciated that the design of a fume cupboard has a marked effect on the rate of air flow; smooth working surfaces, an absence of sharp corners and a profile of sill and jambs which streamline the airflow will increase efficiency considerably. The provision of baffles improves the air flow over the working surface regardless of whether the sash is raised or lowered, and some designs permit air from the room to enter the cupboard near the top to assist with purging (*Figure 1.9*).

Most substances handled in fume cupboards are hazardous in one way or another and many are corrosive. It is therefore

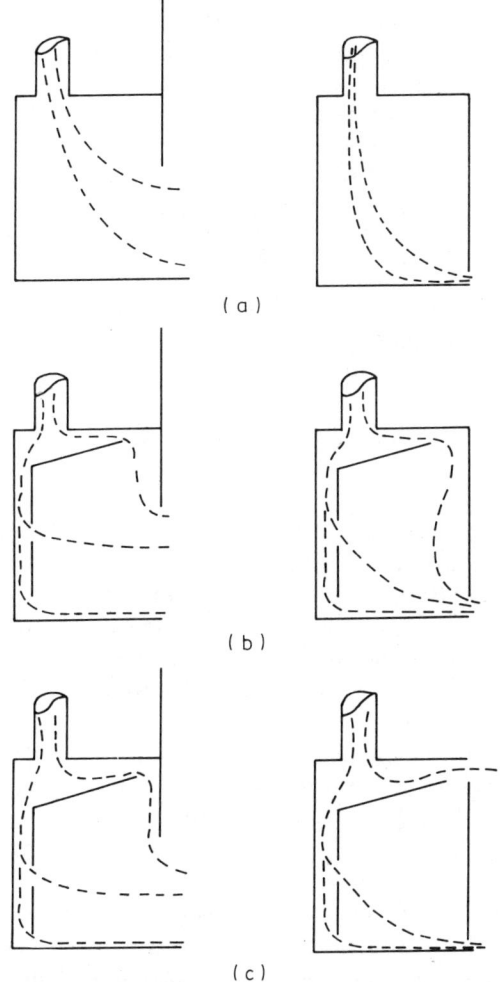

Figure 1.9. Simple fume cupboard: (a) with sash open, draught at the work surface is considerably reduced, whilst with sash closed, air near the top is virtually static; (b) addition of a baffle improves air flow and efficient purging is achieved with sash open or closed; (c) reducing the height of the sash does not affect performance, and with sash closed some air is drawn from the room to dilute exhaust fumes

essential for the units to be constructed of corrosion-resistant materials. Timber is generally unsatisfactory unless it is plastic-coated and various plastic materials such as welded PVC or glass fibre reinforced resins are better. Toughened glass panels have some advantages but it is difficult to seal them effectively. Asbestos working surfaces, even if sealed, are not satisfactory and in view of the health hazard now known to exist, this material should not be used.

The sash must be shatter-resistant as it presents a large area of glass at face level. Toughened glass, wire-reinforced glass or plastic such as polycarbonate are suitable materials and can also be used for side panels where these are required. The sash must be well balanced either with counterweights or spring units, so that it may be opened or closed easily and rapidly in an emergency.

Once a fume cupboard is installed and connected to its ducting it is a major task to move it, and its position must therefore be considered carefully before installation. Siting close to a door should be avoided, because draughts caused by opening and closing the door can cause spillage of fumes; areas where staff circulate should be avoided for the same reason.

Gas, water, drainage, lighting and electrical outlets will be required in the fume cupboard and the location of such outlets and their controls is important. A small sink, sufficient for a condenser tube, is adequate and this should be mounted flush with the working surface so that it may then also be used when washing this surface. Controls must be outside so that they may be operated with the sash closed. Electrical outlets must also be external to the fume cupboard and must be switched units preferably of a flashproof design. Adequate lighting units must be installed and sealed against corrosive vapours and provision made for changing tubes easily (*Figure 1.10*).

The material used for ducting should be selected with care, as it is usually somewhat inaccessible once installed. PVC or polypropylene are satisfactory materials but in the event of fire they will soften and distort. If the temperature rises greatly, PVC produces dense toxic and corrosive fumes. Polypropylene and PVC are perhaps the only available materials suitable for use with perchloric acid. Some Area Fire Authorities require ducts to be fitted with dampers which close automatically on temperature rise — this point should be settled before installation.

Fans should be constructed in such a way that the motor is not situated in the air stream, although for small installations axial fans coated with protective resins are adequate.

Figure 1.10. Fume cupboards. Purpose-built island-sited unit incorporating four fume cupboards with storage facilities in an organic chemical laboratory. Matching laboratory benching, writing areas and bookshelves were also incorporated in this scheme, which was designed and manufactured by A.R. Hoare & Co. Ltd (courtesy A.R. Hoare & Co. Ltd)

For larger installations centrifugal fans are more satisfactory, particularly as the motor can be removed for service without disturbing the ductwork (*Figure 1.11*).

The outlet to atmosphere must be arranged so that fumes cannot re-enter the building. This is best achieved by using an outlet of adequate height above the building. Consideration

Figure 1.11. Fume extraction. Roof area of a university chemical laboratory showing fans and ductwork designed to provide efficient discharge of fumes into the atmosphere clear of roof and building (courtesy A.R. Hoare & Co. Ltd)

Figure 1.12. Fume cupboard and ventilation exhaust showing the fume and ventilation outlet coupled to the system shown in Figure 1.8 (courtesy A.R. Hoare & Co. Ltd)

must also be given to prevailing weather conditions and to the proximity of other buildings which may affect the distribution of fumes in the immediate locality (*Figure 1.12*).

The design of the actual cupboard is an important consideration and varies considerably from one type to another, although they may look very similar. The main variation is in the arrangement of the extract and baffle. The more advanced systems produce a more uniform air flow over the working surface whether the sash is open or closed, but of course one must accept the higher cost of a more sophisticated design.

Most fume cupboards have a working surface at normal bench height but for some purposes floor-standing units may be required. These enable workers to walk into them, and large equipment such as gas cylinders may also be wheeled in and out.

LIGHTING

Modern buildings are usually provided with large window areas in order to take advantage of natural lighting which, in most latitudes, provides sufficient light for much of the working day. However, natural lighting must obviously be supplemented with artificial lighting[4].

Laboratories require a level of illumination of about 600 lux (1 lux = 1 lumen/m^2). Reception areas, libraries and stores need about 200 lux and offices about 400 lux, these levels being generally provided by means of fluorescent tubes fitted with diffusers. The number and wattage of tubes needed depend on several factors, such as the decor of the room, height of fittings above the bench and on the cleanliness of the fittings. Light output from fluorescent tubes decreases with age and an old tube in a dirty fitting can give as little as 50% of its rated output (see Chapter 8). Measurements are taken at the working surface.

Tubes are manufactured to give various colours of light, the 'daylight' types being the most acceptable for laboratories. In areas where accurate observation of colour is necessary, such as in photographic units or paint control laboratories, 'colour matching' tubes are necessary. These give an output approximating to north daylight. In offices and circulating areas, 'warm white' tubes give a less harsh illumination. Warm white and daylight tubes are the most efficient, producing about 65 lumen/W, whereas the colour matching type produces about 42 lumen/W. Even the lower figure is substantially better than tungsten lamps, which give about 14 lumen/W. It is advisable not to mix different types of tube in any room as the effect is distracting.

Inadequate illumination is not only unpleasant but produces severe eye-strain, especially in older persons, and after long periods may result in permanent damage to the eyes. Fine graduations on equipment are difficult to read with poor illumination, and errors can be made and accidents caused. Similarly, lighting which is too bright gives rise to discomfort due to glare; this may cause severe headaches in many people. Gloss paint and shiny working surfaces increase glare, although this can be greatly reduced by the use of diffusers with the lighting units.

Even illumination overall is most desirable; if a working area is alternately dull and bright as one moves about, the eyes cannot accommodate rapidly enough to compensate. Manufacturers recommend maximum spacing to height ratios for their fittings and these figures should not be exceeded. Illumination reduces with the square of the distance between light source and working surface, and the height of the fittings will therefore influence the number of fittings required.

It is important for the ceiling to be well illuminated and fittings should allow a proportion of their light output to be projected in an upward direction. If flush fittings are used this is not possible, but the provision of a light-coloured floor covering will overcome this problem to some extent. Fittings for laboratories should be chosen for their performance rather than by their appearance.

For certain types of work involving fine examination or assembly, very high levels of lighting may be necessary — this can be provided by local high-intensity adjustable lamps. There are models available which are fitted with a magnifying lens, the light source being either tungsten filament lamps or fluorescent tubes. The latter are preferable as they produce practically no heat.

Emergency lighting

In premises which are occupied at night, it may be necessary to provide emergency lighting in certain areas such as staircases, corridors and rooms where essential work is being carried out. In the event of a supply failure, the emergency supply can be switched in manually or, preferably, by automatic means. Generators may be used for this purpose or storage batteries can be utilized. To enable the existing lighting units to be used, batteries are normally coupled to an inverter to provide electric power at the same potential as the normal mains supply. Without this facility, low-voltage lamps will be necessary in addition to the normal lighting.

ELECTRICAL SUPPLIES

The amount of electrical and electronic equipment in laboratories appears to increase continually and in many departments there is a shortage of mains voltage outlets, making the use of multi-way adaptors an undesirable necessity. The cost of providing adequate outlets at the bench is only marginally greater than purchasing adaptors and far more satisfactory. The maximum load likely to be drawn by bench-standing apparatus is about 1 kVA per metre of bench, e.g. a laboratory having 14 m of bench requires about 14 kVA, i.e. 60 A at 240 V. In addition, equipment such as autoclaves, stills, etc., will require separate supplies of suitable rating.

Bench outlets should be fitted with flashproof interlocking switches mounted above the back edge of the bench and distributed along its length in such a way that long trailing leads are not required. They should be wired as a ring main and each laboratory provided with an easily accessible master switch to enable the supply to be switched off in an emergency. Local fuses or overload cutouts should be installed to ensure that a short circuit does not interrupt the supply to other parts of the building.

Three-phase supplies are required for some equipment where large motors or high-power heaters are involved, such as workshop machinery, glass-washing machines and large mixers, and at least one three-phase supply of adequate rating should be available on each floor. There may also be a need for supplies at voltages other than the normal mains supply, and if these are fed to benches, special outlets correctly labelled must be used to prevent the accidental connection of the wrong supply to a particular piece of apparatus. Many electronic devices operate from d.c. voltages of 5–30 V and these may be energized from individual power packs in each laboratory or from one central unit. The former system is preferred, because greater stability is achieved and the problem of voltage drop over long cables is avoided.

Emergency supplies

In the event of a breakdown of the public electricity supply, equipment which must be kept running may be switched, either manually or automatically, to an emergency supply, usually from a generator. The cost of generators rises rapidly with their power output capability and, with few exceptions, it is not feasible to supply the entire building. It is good practice to connect selected outlets to the emergency supply as a feed for equipment which must be kept running at all times. Refrigerators and freezers will normally hold their temperature for several hours if not opened, and unless prolonged power failures are anticipated they do not require standby power.

GAS SUPPLIES

In the past, a large amount of gas was used in laboratories for equipment such as ovens, stills and incubators, but today the traditional bunsen burner is probably the only gas appliance in common use. As there does not appear to be a totally satisfactory substitute for this basic service, most laboratories will require a gas supply, except in those areas where a high fire risk exists. In such areas, gas should not be installed, even if it is proposed to blank off the pipes.

All outlets should be of a safe construction with levers which lock in the 'off' position. A very small leak can cause a build-up of gas to an explosive level in a room which is left closed for a long period such as a weekend. If double outlets are spaced along the bench about 2 m apart, the use of long connecting tubes is avoided. Some laboratories have master stopcocks, clearly marked, outside the rooms so that they can be turned off in the event of a leak or fire.

WATER SUPPLIES

Adequate supplies of cold water must be available in all laboratories; one outlet per 3 m of bench is sufficient for the majority of situations. A teaching laboratory catering for a large number of students will probably require twice this number. Since the supply is normally taken from storage tanks at roof level, the pressure varies with head of water. This problem may be overcome by using a booster pump to raise the pressure to about 3 kgf/cm^2, which is satisfactory if water jet suction pumps are to be used. Some local authorities will not permit the use of these pumps, however, owing to their high water consumption. High-pressure outlets should be colour coded for easy recognition. Several types of outlet are available and these should be chosen according to their function. Most will be suitable for connection to flexible tubes for use with condensers or similar apparatus, but at least one tap in each laboratory should be of a larger bore because large volumes of water may be required quickly in an emergency such as an acid spillage.

In general, the demand for hot water in laboratories is fairly small but is useful for warming solutions, hand washing, etc., and at least one hot tap should be provided in each laboratory. Mixer taps are well worth the small extra cost. On no account should gas water heaters be installed in laboratories, but if a hot water supply is not available an electric storage-type water heater is a satisfactory and inexpensive alternative.

Drainage

The very large sinks common in laboratories a few years ago are giving way to smaller units or drip cups. One function of sinks in the laboratory is to carry away the cooling water, or the outlets from water jet pumps, and for the disposal of small amounts of waste materials. Large sinks are therefore not necessary for these purposes.

Sinks are made of various materials, glazed fireclay being the most satisfactory for the majority of applications. Stainless steel is also used but some grades are of dubious quality and need to be chosen with care. Polythene, polypropylene and glass fibre sinks are easily scratched and difficult to keep clean. Whichever type is chosen, it is important to ensure that the joint between bench top and sink is watertight and will remain so under all conditions.

Waste pipes and fittings are an important feature, because replacement of an unsatisfactory system is a costly and disruptive exercise. Even with the greatest care, corrosive materials do find their way into the waste pipes and not only damage the system but cause further damage to the building because of eventual leakage which may go undetected for long periods.

Plastics materials, such as polythene, are very satisfactory provided they are of good quality and correctly fitted. Long horizontal pipe runs tend to sag and must be properly supported. Demountable joints and cleaning eyes are necessary to enable blocked pipes to be cleared. In some laboratories, large catch pots are used to take waste from several sinks but it is better to install individual traps for each sink. The design should be such that the trap can be dismantled without the use of tools to enable blockages to be cleared quickly and small items such as magnetic stirring bars to be retrieved.

In laboratories where highly toxic or radioactive materials are in use, special consideration must be given to the waste system (see Chapter 7). Glass or glass-lined iron traps and pipelines are recommended when working with highly corrosive liquids. Although costly to install, they have a very long life and require the minimum of attention[3].

Distilled and de-ionized water supplies

Distilled water may be supplied to individual laboratories or to supply points on each floor from a central still. Alternatively, individual stills can be installed in those rooms where a supply is needed. Where demand is very large, the former system may be preferable but it has some disadvantages. Pipework and storage tanks of a suitable nature are expensive and can give rise to contamination of the whole supply if corrosion, leakage or algal growth develop. Water lying in pipes for long periods may leach soluble materials from the glass or metal or from jointing materials. When the still is shut down for repairs or maintenance, the entire building is without a supply. The capital outlay for a large still, together with the provision of power supplies, cooling water, etc., is very considerable although the output in terms of l/kVA may be higher depending on the design.

Small stills producing 4—8 l/h are relatively inexpensive, take up little space, can be easily maintained and may be shut down individually for servicing. Whilst gas-heated units are available, electric heating is more satisfactory and produces about 1.3 l/h per kVA. Regular descaling is necessary to achieve this output, particularly in hard water areas, and it is advisable to select a design which can be descaled with the minimum of dismantling. With some models no dismantling whatever is required.

If live steam, free of volatile impurities, is available in the building, strip-action stills may be used to advantage. These consist of a condenser into which the steam is passed via a system of baffles to strip out major contaminants which are flushed away with the cooling water. Such units are virtually maintenance free, take up minimum space and have a high output.

Storage of distilled water should be in reservoirs of either glass, PVC, polythene or polypropylene. These may be fitted with float switches which cut off the still when full and restart it when water is drawn from the reservoir.

High-purity water may be produced by de-ionization with resins formulated for this purpose. Anionic and cationic resins may be used in separate columns, mains water being passed through each in turn; alternatively, a mixed resin can be used in a single column. In the former case, the resins may be regenerated by the user, while in the latter a new resin bed is installed when required, the exhausted one being returned to the supplier on an exchange basis. Units are available in various sizes from small bench units up to large installations providing several hundred gallons per day.

In soft water areas, the amount of de-ionized water produced can be very large before the resin becomes exhausted, but in hard water areas the output is considerably reduced. It is often more economical to distil the supply water first and to pass this directly from the still to the de-ionizer and thence to a storage reservoir. Whichever system is used, it is essential to monitor the quality of the output by measuring its conductivity. Despite the cost of the resin, de-ionizing can be an economical way of producing high-purity water because, unlike the still, no heating is required. The actual cost will depend on the concentration of dissolved impurities present in the supply water, which will vary from one area to another.

Steam supply

Certain laboratories will need a piped steam supply as a service for heating, operating autoclaves, etc., or for conversion to distilled water. It is costly to install, but if a high-pressure boiler is used for space heating, the additional cost is not great. All pipework must be of a high standard and properly lagged to prevent heat loss and to avoid accidental burning. Open-ended steam valves should not be fitted to benches as they are a hazard, and steam should only be used on apparatus to which it can be directly plumbed. Reducing valves must be fitted in the supply lines at the appropriate points, and only apparatus designed to withstand the supply pressure and which are fitted with reliable safety valves may be used.

PIPED GASES

Gases which are used routinely may be piped at reduced pressure (say 4 kgf/cm^2) to suitable outlets in the laboratories from a central cylinder store. Only gases which are non-flammable and non-toxic should be supplied in this way. Compressed air, nitrogen, argon, etc., may for example be piped to laboratories engaged in gas chromatography, but on no account should such gases as acetylene, hydrogen and chlorine be supplied by this method unless the pipework is installed outside the building.

Outlets at the bench must be fitted with pressure-reducing or flow-control regulators, clearly labelled and regularly maintained. Oil or grease must not be allowed to come into contact with them and readily accessible shut-off valves must be installed outside the laboratory for use in emergencies.

Safety in design

At all stages of design, safety is of paramount importance. In most countries, building legislation will impose certain design requirements but little if any guidance is given regarding details such as floor coverings and lighting of stairways. Each room should, wherever possible, have more than one exit which, at ground floor level, may be an opening window of sufficient size. In small rooms where two doors would take up valuable space or otherwise be impractical, a 'kick-out' may be incorporated in one of the walls, preferably a corridor wall. This consists of a hole covered with a light panel which is easily broken down and of sufficient size to allow a large adult to pass through. It must, of course, be clearly labelled and unobstructed at all times on both sides.

The hazards of fire are referred to in Chapter 7 and corridors and staircases must be provided with self-closing doors to contain smoke and reduce draughts which would encourage the spread of fire.

Apparatus producing ionizing radiations, such as X-ray equipment, must be housed in rooms which are screened to prevent radiations passing to other areas. Such screening (usually lead panels, lead-loaded concrete or barium-loaded bricks or plaster) is extremely heavy and very costly and allowances should be made for its inclusion where necessary.

Decoration

Decoration serves a number of purposes apart from providing a pleasant environment. It has a marked effect on the level of lighting (particularly daylight utilization) and can reduce the cost of artificial lighting. It preserves the structural materials of the building and provides surfaces which are easily cleaned and disinfected.

For general laboratory use, light colours are preferred for ceilings, walls and floor coverings. Gloss paints tend to produce glare, so that matt or eggshell finishes are more satisfactory. In areas where hazardous materials such as radio-isotopes and bacterial cultures are handled, gloss paints may be necessary as decontamination is more easily carried out.

Timber, apart from bench surfaces, may be gloss painted or treated with varnish.

Plastic laminates make excellent decorative finishes in areas where dirty material is handled, such as in animal houses; although expensive, such laminates require little or no maintenance. Ceramic tiles can also be used to advantage in these areas.

Decoration in offices, common rooms, lecture rooms, etc., is not so restricted and the careful use of colours can be very pleasing. As the decor is not subjected to such harsh treatment as it is in laboratories there is no reason why wall coverings or emulsion paints should not be used. Where an office is occupied by one person, he may be given the choice of colours; if a number of persons use the room, however, the scheme should be chosen by mutual agreement. There is no doubt that people's temperaments are affected by the decor of their place of work and they can become inefficient and even aggressive in surroundings which they find depressing or otherwise unpleasant.

Allocation of floors in multi-storey buildings

Deciding which rooms should be sited on the various floors of a building is not such a simple exercise as it might at first appear. Incoming goods must be delivered and waste material disposed of; hence there is a good case for siting departments with a large throughput on the ground floor. It can also be argued that laboratories using heavy equipment should be on the ground floor. Where flammable materials are handled, the risk of fire spreading is reduced if these laboratories are situated on the top floor. One is soon forced to the conclusion that nearly all departments should be on the ground floor and, indeed, if a large site were available, a single-storey building has a great many practical advantages.

There is seldom a good case for using basements as working areas. Boiler plant, ventilation systems, long-term storage facilities and so on can be accommodated below ground level without condemning staff to work in depressing environments.

The ground floor forms a natural reception area with offices, meeting rooms, cloakrooms, etc., adjoining. If sufficient space is available, stores (with access for delivery vehicles) should also be at ground level, perhaps at the rear of the building (see Chapter 4).

Workshops and laboratories using heavy apparatus are also best placed at ground level, with remaining laboratories on the upper floors. Animal accommodation is usually on the top floor as it is easier to isolate these rooms, although waste disposal can present some difficulties (see Chapter 6).

Service rooms which are shared by all departments, such as glassware washing or photographic facilities, are best positioned near the centre of the building to reduce to a minimum the distance between them and their users.

Rooms which are devoid of daylight are occasionally to be found in the central core of large complexes. Such areas are ideal for storage, photographic dark rooms, toilets, etc., but should not be used as normal working areas where staff would be expected to remain for long periods of the day.

REFERENCES

1. Nuffield Foundation, Division for Architectural Studies, *The Design of Research Laboratories,* Oxford University Press (1961)
2. BS 3202: 1959, *Recommendations on Laboratory Furniture and Fittings.* British Standards Institution, London
3. Hughes, D. and Cullingworth, R., 'Laboratory fittings and waste systems', *Chemistry in Britain,* 8, 470–474 (1972)
4. *Lighting in Offices, Shops and Railway Premises,* Booklet No. 39, Health and Safety Executive; HMSO, London (1976)

FURTHER READING

Adaptable Furniture and Services for Education and Science, Paper No. 6, Laboratories Investigation Unit; Department of Education and Science, London (1972)

32 Laboratory planning and layout

Adaptable Laboratories: Practical Observations on Design and Installation, Paper No. 7, Laboratories Investigation Unit; Department of Education and Science, London (1974)

An Approach to Laboratory Building, Paper No. 1, Laboratories Investigation Unit; Department of Education and Science, London (1969)

Bristol Polytechnic, The Charles Darwin Building. *Laboratory Investigation,* Unit Paper No. 9, Department of Education and Science; HMSO, London (1977)

BS 1710: 1960, *Specification for the Identification of Pipelines.* British Standards Institution, London

Ferguson, W.R., *Practical Laboratory Planning.* Applied Science Publishers, London (1973)

Munce, J.F., *Laboratory Planning.* Butterworths, London (1962)

Research Laboratories, Design for Flexibility, Paper No. 10, Laboratories Investigation Unit; Department of Education and Science, London (1977)

The Conversion of Buildings for Science and Technology, Paper No. 8, Part 2, Laboratories Investigation Unit; Department of Education and Science, London (1977)

The Ventilation of Buildings, Fresh Air Requirements, Technical Data Note No. 19, Health and Safety Executive; HMSO, London (1976)

2
Selection and Management of Staff

Without doubt, one of the principal resources of any organization engaged in the provision of goods or services is the staff that are employed in providing them. With the increasing sophistication of modern laboratory equipment, one may be tempted to think that the provision of elaborate apparatus or costly accommodation is more important than the recruitment and retention of efficient staff. This is not so. Staff who are properly qualified, selected, trained and paid for the tasks which they are expected to perform are much more likely to overcome, by extra effort or innovation, the difficulties caused by the temporary shortage of equipment or financial resources that are encountered in all laboratories from time to time. Those organizations which have developed good staff selection and management techniques, and are aware of the importance of career development and good labour relations for their staff, are far better equipped to surmount difficulties that would cripple their less-concerned competitors.

Job description

Before any post is advertised, it is important to define the job that has to be done, so that a person with the most appropriate qualifications, experience and personal qualities, such as motivation and the ability to co-operate with immediate colleagues, may be recruited.

Accurate and detailed job descriptions are essential for most, if not all, posts. In some countries there may be a legal obligation to provide all employees with a copy of their job description in connection with their contract of employment[1]. Quite irrespective of such legal obligations, these job descriptions will be found useful both to the employer and to the employee. Job descriptions are used by management as a background to three undertakings. First, to decide the appropriate grade and salary level for a particular post; secondly, in the composition of an advertisement describing the vacancy; and thirdly, to assist the interviewer when selecting the most appropriate candidate to fill the post.

Copies of the job description are usually supplied to all applicants at an early stage in the negotiations, so that in the event of one of them being accepted, he will know exactly what is to be expected. This procedure will also enable the candidate to approach the interview in a more informal manner, so that he will be able to satisfy himself that he really wants the job and will be happy doing it. For relatively junior posts, the description may need to go into considerable detail, whilst for the more senior posts the description may be couched in broader, more general terms. Job descriptions should give details of all the work to be carried out, together with the duties and responsibilities involved. They should not only mention the person to whom the successful applicant will be responsible but also enumerate the persons for whose actions he will be held responsible.

It does not necessarily follow that because there is a vacancy, it must be filled. In an example known to the authors, efforts were made to recruit by advertisement for a post that had been vacant for over one year. Throughout the whole of this period, no staffing difficulties manifested themselves and it obviously did not occur to those responsible for the implementation of recruitment policy that there would be little need to refill the post at all.

Job descriptions are best compiled by persons not too remote from the job that is to be filled. In large organizations, the staff of the personnel department will have a major part to play in this because they will have the experience that will ensure a degree of continuity. Generally, the person best able to write an accurate account of the tasks to be performed and to suggest the qualities most likely to be needed in an individual to fill the vacancy, is the immediate superior or line manager responsible for the section or department concerned, but he may need the help and guidance of the professionals from the personnel department before the description is completed.

The advertisement

Past experience is generally the best guide in reaching a decision as to when, where and how a particular vacancy should be advertised.

For jobs that require bright school-leavers or newly qualified graduates, there is little point in placing expensive advertisements in August or September because most of the really bright job-seekers will already have been engaged by firms who had the foresight to start their recruitment campaigns earlier in the year. Similarly, many organizations find that they obtain a poor response from potential recruits if the advertisement appears in a national newspaper on the first few days of the working week. Many junior staff scan the Friday, Saturday and Sunday papers for suitable job vacancies and write their applications immediately. Where to advertise is important in as much as certain newspapers, journals and technical magazines tend to specialize in advertisements for particular types or classifications of scientific staff. It should be remembered that a few lines in a national newspaper may cost more than a half-page display in a journal with a limited circulation.

In general, advertisements are placed directly with the newspaper or journal concerned but they may also be placed through advertising agencies, generally at no extra cost. This can prove very useful, because many agencies will give advice on the form, wording and placing of an advertisement without charge; the agency's income is by a commission or discount from the publication concerned. A number of organizations recruit their senior staff with the help of management selection agencies which advertise on their behalf and provide a shortlist of selected candidates for interview. This is a fairly trouble-free approach to recruitment in cases where the organization does not have staff with sufficient expertise and experience to deal with the whole operation, but nevertheless this type of recruitment may prove relatively expensive.

Some firms and organizations have entered into agreements with their staff that vacancies for other than the most basic grades will first be notified to the existing employees before they are advertised externally, whereas in other organizations it has become the practice to fill all senior posts from within. Clearly, this practice has important advantages so far as career development for existing staff is concerned, but there may also be disadvantages. If the process is carried too far, it is possible that the organization will become inbred and stagnant and lose the drive and ability to perform and compete in the modern world.

Application forms

It is common practice for organizations to ask that persons interested in a particular post which has been advertised, or the details of which have been circulated, should complete an application form. This is usually sent to the applicant, together with the job description and perhaps a general statement concerning the work of the organization. A less formal approach adopted by other organizations is to ask the candidate to write an application giving details of his qualifications and relevant experience. The completion of a standard form will ensure that all applicants will answer the same questions, so that an accurate comparison may be made when drawing up a shortlist. The second approach, however, will generally require the candidate to 'sell himself', but it will not necessarily yield the same information from which to make a first selection.

Sometimes both systems are combined, i.e. the candidate makes a written application, is shortlisted and the personnel officer completes a standard form when the candidate is called for interview. To assess the relative merits of candidates without the aid of a completed application form requires much experience and a fair insight into human behaviour. In a letter, the candidate may gloss over certain aspects of his professional background without being exactly untruthful, but if he is specifically asked to list his previous jobs in chronological order, with dates, he cannot easily avoid quoting the whole of his previous work history without being dishonest. After considerable experience, one is able to detect the more obvious half-truths and evasions without difficulty.

The majority of application forms ask the candidate for details of his medical history, but allow him to elect to discuss this only with the organization's medical officer or adviser. This is an important point that needs to be carefully considered before making an appointment. Not only is the organization interested in recruiting a person who will be fit and able to give reliable service, but it will also be interested in protecting its pension fund from unacceptable or unnecessary burdens. Laboratories are not particularly dangerous places of work and we all have a duty to care for our less fortunate fellows, but reasonable steps must be taken to ensure that any candidate with a history of chronic illness will not become a hazard either to himself or to his colleagues. If there is any doubt, the organization's medical advisers must be consulted. The express permission of the candidate must be obtained in writing before any approach is made to his medical advisers.

REFERENCES

Application forms will require the candidate to supply the names and addresses of at least two organizations or persons to whom reference may be made concerning his reputation and previous performance. Generally, references are taken up once the candidate has been shortlisted, but occasionally the candidate may ask that his present employer shall not be approached unless he has a reasonable chance of being offered an appointment. This is a perfectly acceptable request.

The writing of references is a time-consuming and sometimes difficult procedure which may involve the referee in a great deal of work. It is important in the interest of the business community generally, and of the referee in particular, that his time should not be unnecessarily wasted.

Many organizations use their own standard printed form when taking up references. The form is headed by a short note in which is typed the applicant's name, together with details of the post for which he has applied. The referee is asked to complete the form by giving the answers to a number of specific questions concerning the applicant's qualifications, ability and record whilst working for him or during the period that the applicant was known by him. To some extent this procedure, although it is rather impersonal, makes the task easier for the referee, since he has only to write a few words or a short sentence in answer to each question and nothing of importance is likely to be inadvertently omitted. The deliberate omission, however, will be immediately obvious to the recipient. Sometimes that which is left out is more important than what is actually written. A classic story illustrating this point concerns the referee who, after considering for some time what good he could say of a person who had given his name to a potential employer, wrote: 'I have known A.N. Other for thirty years.' This story is capped by another in which a dissatisfied employer wrote: 'J. Smith has worked for me for some ten years, during which time he has performed his duties entirely to his own satisfaction.'

The taking up of references may be of limited value and, at best, the results may be somewhat doubtful. It is often very difficult to avoid subjective judgements when composing or considering them. There is a duty to be truthful, not only to the person who is being written about but to the organization seeking the information. Bias both for and against the individual may be found; some people are very reluctant to write ill of any person, whilst others may only see his weaknesses rather than his strengths, knowledge of the latter being generally more important. It is not unknown for a difficult or otherwise unsatisfactory employee to be given a rather better reference than he may deserve, simply because his present employer will be more than happy if he obtains another job elsewhere.

On receiving a doubtful or unsatisfactory reference a few organizations will arrange for their personnel officer to telephone the referee for amplification of particular points in the reference; it may be found that he will be much more forthcoming about an individual when speaking on the telephone than he would otherwise be in a letter. Nevertheless, these remarks must also be interpreted with caution, because there will be no legally acceptable proof of what was said. The process of seeking further information concerning an individual by telephone will generally be found more useful when trying to establish the answers to specific points or removing ambiguities or apparent misunderstandings in the original reference.

Interviewing and selection

The purpose of holding an interviewing session is to select from amongst those candidates who have been previously shortlisted, either the most suitable candidate who is then offered the job and accepts, or to prepare, as a result of the observations made during the interviews, a further shortlist where the candidates are ranked in order of suitability from which the final choice will be made. Whichever course is chosen, the final objectives are the same. They are to find the candidate whose qualifications, experience, personality, etc., best match the qualities called for in the job description, and to ensure so far as possible that the job and the organization are suitable for the candidate. These factors must be considered of equal importance and clearly there is little point in offering a post to a suitably qualified person who is likely to become so unhappy in the performance of his daily tasks that after a relatively short period he resigns and the whole expensive recruitment process has to start again.

Interviewing and personnel selection are skilled techniques that need to be considered with some care if the major pitfalls are to be avoided[2]. In fact, the whole process of staff recruitment should be regarded to some extent as a public relations exercise. It is important that the candidate, whether or not he has been successful, should be left with the impression that not only has his application been fairly considered but that he has been received and treated with courtesy.

As far as the candidate is concerned, the interview is likely to fall into two distinct areas — the administrative and the technical. The administrative part is concerned with the checking of qualifications, present rate of pay, notice required to terminate present job, pay scales, leave entitlement, pension schemes and their transferability and possible starting date, etc. This is generally handled by a member of the personnel or administrative staff. The other part of the interview is concerned with the technical details of the post and with the technical suitability of the candidate to fill it. Usually this is dealt with by a person who exercises a supervisory responsibility for the post.

Any interview ought to be regarded as a two-way process. The candidate should be allowed time to collect his thoughts so that he may ask questions or seek clarification of points as they arise, as well as answer the questions put to him by the interviewer. Often, for convenience, the two parts of the interview are allowed to run together. Generally, two interviewers, working together to a prearranged plan, are sufficient. Large panels or selection boards where the candidate is surrounded by a mass of strange and often unidentified faces should be avoided for all but the most important posts. If the interview is concerned with the appointment of a managing director who is expected to take over the leadership of an ailing company, it may well be necessary for him to demonstrate his ability to stand up to questioning under stress, but for the vast majority of posts this is not a necessary or very useful attribute.

Many candidates arrive at an interview in a state of considerable tension and quite often the more they want the job, the greater is their apparent anxiety. One of the objectives of the interview is to establish the facts concerning the candidate's past record and performance and to assess how he is likely to perform if he is given the job. This cannot be established with any degree of accuracy or fairness in an unduly formal atmosphere or if the candidate is unduly stressed. One very effective way of reducing tension is to encourage the candidate to talk by asking questions that need more than a straight 'yes' or 'no' for an answer. Often, this can be achieved very simply by rephrasing an intended question. If he is asked 'Did you enjoy your time at college?' he will probably give a yes or no answer, but if the question is reframed as 'Tell us about your life at college' it will almost certainly elicit a much more informative reply. The importance of creating an atmosphere free from tension and stress cannot be overemphasized; both authors have had personal experience of applying for posts where, because there was no very great pressure on them to be successful, they were relaxed and evidently gave such an impression of professional competence that they were asked, at the end of the interview, to accept the job.

It is essential that before the candidate is seen by the interviewers there should be time for a few minutes' discussion concerning his papers and references, etc., so that any special points that need to be cleared up or focused upon may be agreed and allocated to a specific interviewer to raise at the appropriate time.

The interviewer or, if there are more than one, the person who is to control the interview, should start the proceedings by welcoming the candidate in a warm and friendly manner and he should then introduce himself together with any other person present, mentioning their name and position in the organization. Every effort must be made to ensure that the interview is conducted in an efficient but relaxed atmosphere. When the interviewer has obtained the answers to his questions, he should pass the initiative to one of his colleagues by means of a prearranged signal so that the discussion will flow smoothly without awkward gaps. When all the points have been established and the candidate's own questions answered, the interview should be closed, usually by informing him when he may expect to hear if his application has been successful or not. He should be thanked for attending the interview and passed on to someone who will see that his travelling expenses are reimbursed before he leaves.

After the close of the interview, the interviewers should take a few minutes to discuss the candidate's answers and performance and to award a rating, so that he may be compared in an objective manner with other candidates.

At some stage, either before or after the interview, it will be necessary for the candidate to be given an opportunity of seeing the laboratory or department concerned, so that he will be able to make his own assessment of the environment in which he will be expected to work. This will also provide an opportunity to introduce him to some of his potential colleagues whose attitudes and demeanour will be important to him when deciding whether or not to accept the job if it is offered. Many organizations arrange for all shortlisted candidates to inspect the laboratories before they are interviewed; others restrict viewing to those among the shortlisted who have a reasonable chance of being offered the post. Much depends on the number of candidates and the particular circumstances in which the vacancy is being filled. It may be considered that it is a waste of time and executive effort to show a number of recruits around the laboratories when only one of them will ultimately be employed. Against this may be set the advantages accruing when the occasion is made the subject of a public relations exercise.

Apart from the formal type of interview described above, many organizations are prepared to arrange informal sessions during which prospective applicants may meet the immediate supervisor individually as well as other members of staff in order to discuss the work, usually at the place where it is carried out. This particular approach to selection is to be commended and it is unfortunate that it is not more widely adopted. Not only does it allow the more retiring candidate to produce a better effect and show himself to his best advantage, but because of the informal and relaxed atmosphere it gives the interviewer an opportunity to observe the candidate and his behaviour in a stress-free environment.

FINAL SELECTION

Assuming that there have been a large number of applications for a particular post which has been advertised, and that a number of persons have been shortlisted and interviewed, there now remains the task of choosing the candidate who is to be offered the post. At this time it is usual to select one or two to act as reserves should the first choice for any reason fail to accept the offer. Sometimes, if the appointment is to be made at a high level or there are special difficulties over, for example, the salary or starting date, it may be necessary to consult other managerial staff before a decision or offer of employment can be made.

Occasionally, none of the persons interviewed will be found to reach the required standard and the whole process will have to be started again. Before this is done, the persons who initiated recruitment should consider very carefully why the first attempt failed to meet requirements. Was the timing or placing of the advertisement inappropriate? Was the salary offered too low? Were there some other undesirable features attached to the job? The timing or placing of an advertisement can easily be modified, but the raising of a salary or staff grading may require a great deal of thought, not the least being their effects on other staff at present in post. It may be possible to reconsider the job description to check whether it can be modified to remove an undesirable feature, but it is essential that such features are not so glossed over before the advertisement reappears in the press as to amount to deception.

If the interview and selection process has been protracted, which unfortunately is frequently the case, it will be necessary to select more than one candidate, usually two or three. It may be found that the first choice, on being made an offer, declines to accept because he has responded to market forces and accepted an earlier or better offer from another organization. Many candidates, especially school-leavers or newly qualified graduates, apply for more than one job at a given time and subsequently accept the first good offer that they receive. This is a not unreasonable attitude, particularly when suitable posts are difficult to obtain. Candidates may also decline an offer because they have decided, on reflection after the interview, that the post offered was unsuitable; this is regrettable but has to be accepted with equanimity. To avoid this kind of situation, the negotiations should be carried through with reasonable speed. To have one's first choice decline an offer is bad enough but to find that the second and third choices have meanwhile found posts elsewhere can be particularly disconcerting.

Some interviewers make a practice of asking candidates with whom they have been impressed if they would be likely to accept an offer if it were made. At best this is a somewhat dubious practice since to some extent it prejudges the candidates who have yet to be seen and may therefore lead to embarrassment later. In addition, it may raise false hopes, cause an applicant to reject an otherwise suitable offer from another company and leave him jobless.

What are the factors that have to be considered when making the final selection? Depending on the precise nature of the job, as defined in the job description, they are (a) qualifications, (b) work experience, (c) references, (d) personality, (e) health, and (f) social and other leisure activities. All of these will have to be assessed for each candidate so that his attributes may be weighed and compared with the others and the job description. Rarely will it be found that any particular candidate will possess all the necessary qualities but some effort must be made to balance the essential and desirable qualities against the inevitable shortcomings in a particular individual.

When making the final section, the danger of appointing a person who is overqualified should not be overlooked. At the present time, the market in some fields is over-subscribed with graduates who, desperate for work, will apply for any job for which they consider they are remotely suited. Advertisements for posts which would normally be filled at technician or junior technical officer level by persons holding technical or technological qualifications are now attracting replies from large numbers of newly qualified graduates unable to obtain posts appropriate to their qualifications. Many such appointments, which at first sight might appear to be doomed to failure, have been made with considerable success, but the practice of making such appointments is not without its pitfalls. Generally, newly qualified graduates will be lacking in experience and some of the practical ability that one would normally expect to find in a traditionally trained technician. This situation will have to be remedied by the provision of rather more in-house training than would otherwise be the case. There is also the possibility that the overqualified appointee will become so bored, frustrated or disillusioned in performing tasks that he comes to consider to be beneath him, that it will lead to his premature resignation.

It may be expected that when the present rather slack industrial situation improves, as it inevitably will, the many graduates who have taken such jobs will look for better posts elsewhere. Nevertheless, a modest degree of overqualification may be tolerated if it can be seen and explained at the interview that there is a clear path by which the candidate can progress at a later date to another more suitable post.

It is the practice of many organizations to keep records of the assessments made at interview of all applicants, together with notes of the reasons for rejection. It is not unknown for an unsuccessful candidate to allege that he or she has been discriminated against on grounds of colour, race, creed, trade union activity, or sex[3,4]. In a number of countries (including the UK) discrimination on one or more of these grounds is illegal. It is worth noting that reference to such records could be used to substantiate the defence should a legal action be brought against the employer.

Contracts and conditions of service

When the most suitable candidate has been selected, a formal offer of appointment must be made. Occasionally, before the contract (which is a formal legal document) is drawn up the successful candidate will be informed, by means of a letter which sets out the broad general conditions attached to the offer, that he is to be offered the appointment. If the conditions are accepted the contract is drawn up and sent to him, together with the organization's general conditions of service, for approval and signature. More generally, he is informed by means of a brief letter that he has been successful, and two copies of the contract are enclosed, one of which he is expected to sign and return, thereby signifying his agreement and acceptance. There are a number of major and minor variations to this procedure, but in every case the end result is the same, i.e. each party to the contract is legally bound to observe its conditions[1]. It is rare for an employer to sue an employee for breach of contract if only because his chances of being awarded damages are very small, but the reverse is by no means so. In general, the terms of a contract may not be varied by one party without the consent of the other.

Apart from the names and addresses of the employer and the employee, it is usual to include only the most essential information and conditions, etc., on the contract document, such as:

(1) Nature of the post together with the rank or grade if appropriate, e.g. technician, senior technician, technical officer, scientific officer, etc.
(2) Name of the laboratory, section or particular establishment where the person will be employed.
(3) Salary or salary scale, together with the starting point.
(4) Hours of work.

(5) Leave entitlement.
(6) Date of commencement.
(7) Date of termination if the contract is for a fixed term only.
(8) Length of probationary service before the appointment is made permanent.
(9) Whether the employee is required to join a pension scheme and whether the scheme is contributory or non-contributory.
(10) Length of notice required by either side to terminate the contract.

Other information or conditions of service which may apply to all or large groups of employees in the organization are usually conveyed to the prospective employee by means of a printed booklet which sets out the general conditions in a slightly less formal manner than is used in the contract. These general conditions of service may be varied from time to time by agreement between the employer and the trades unions or other representative organizations. If this is so, then the contract may make reference to this general provision.

Many employers make offers of employment where the contract is offered and signed by the employer subject to certain specified conditions. The most common condition imposed in these circumstances is that the employee shall be examined by the employer's medical representative to ensure that he is medically fit and therefore will not be an undue liability to the pension fund. Another such condition may be that the post must be taken up by a given date.

Induction

Induction may be defined as the process of introducing the new recruit to the organization, its staff and its systems of work so that he may be harmoniously integrated and become productive and efficient at his job in the shortest possible time. Almost every reader will be aware of the bewilderment, anxiety and doubts that are too frequently encountered in the first few hours or days of taking up a new job. Too often the so-called 'induction' consists of handing over one's income tax and insurance papers, drawing one's white coat from the stores, being introduced to Mr Bloggs, Harry and Tom and then being left to get on with it until lunchtime.

At the other extreme the recruit is taken on a guided tour of the whole plant or institute and introduced to dozens of staff, many of whom he may never see again (and most of whose names he has no hope of remembering). He is then passed on to a succession of other guides who proceed to overwhelm him with technical and other information.

Induction is really an exercise in communication. Some information is necessary to the recruit before he arrives, some can only be given when he arrives and the rest may be given over the ensuing few weeks or months. The person responsible for recruitment and induction (which is part of the recruitment process) should consider the following factors when making plans to introduce the new arrival to his work and environment:

(1) What information does the recruit require?
(2) Who is best suited to give him this information?
(3) When should the information be given?

For convenience, the information to be given may be divided into at least three groups.

(a) Personal information — where may he leave his outdoor clothes and keep his personal possessions; where are the toilet and washing facilities; when and where may he obtain refreshment; when, where and how is he to be paid, etc.
(b) Job information — what precisely is he to do and how and where is he to do it; which particular tools and equipment is he to use; how are they to be used and who looks after them; what happens when they go wrong.
(c) Staff information — who is he responsible to; who is he responsible for; who are his immediate and remote colleagues; when and where can they all be found.

It is essential, no matter in what order the information is imparted, that specific persons are charged with giving each item of information; they must also be required to report that the information has been given and absorbed. Many organizations arrange that, for the first few days or weeks, the recruit shall work with a colleague of equal rank who is able to resolve his more immediate problems. The colleague ensures that the recruit is sympathetically introduced to his work and responsibilities and he generally acts as a counsellor during the induction period. This process, although it may seem expensive, has important practical and social advantages both to the individual and to the organization.

Training and further education

A common dictionary definition of training is 'practical education in any profession, art or handicraft'. In its basic form, as applied to laboratory work, it is concerned with the development of skills to improve the performance of individuals or groups of individuals so that the work may be carried out with greater accuracy or more efficiency; thus it has an important role to play in that part of laboratory organization concerned with the reduction of costs.

In the field of laboratory work, training is inescapably linked with academic and technical education. Training to improve performance may be regarded as a continuation of the induction process to which all new entrants have, to a greater or lesser extent, to be exposed depending on their qualifications and previous experience. The induction process is concerned with adapting the recruit to the organization and its work as quickly and efficiently as possible, but in the wider context training is concerned not only with the corporate needs of the organization but with the developing needs of the individual. Training is a communication process and, if it is to be successful, the needs and abilities of the learner must be considered at all stages so that he is encouraged to co-operate. In this manner, both he and the organization will derive the greatest benefit.

Whilst training is concerned with improving performance in the present job, education is generally concerned with the acquisition of knowledge that may lead to a better job or increased responsibility in the present one. If the organization has a realistic career development programme, linked with an awareness of its probable staffing requirements, the effort put into training will rarely be wasted when the trainee looks outside the organization to find his better job. It is always disappointing when a person who has been trained, often at considerable expense, leaves to take up a post with another organization. It should be remembered that some benefit will have been obtained whilst that person was under instruction and the outstanding loss will be balanced in part by the occasional recruitment of staff who have been trained by other organizations. Cross-fertilization of ideas in the build-up of an efficient labour force can be just as beneficial in laboratory work as it is usually considered to be in the strictly biological sense.

Training may be carried out either internally or externally. At its simplest, the trainee is merely shown how to carry out a particular task, and it usually takes the form of a demonstration at the bench followed by practice under the supervision of a person with the necessary experience and ability to convey the information. This form of 'on the job' training is sometimes rather disparagingly referred to as 'standing beside Nellie', but its advantages should not be underestimated. Not the least of these is the opportunity for person-to-person contact in the introduction of new ideas or novel systems of work. Where more complex operations which will require the absorption of theoretical background information are to be introduced to the trainee, it may be necessary to arrange external 'off the job' training. This is usually carried out at technical colleges or at equipment manufacturers' or suppliers' training schools. Advanced level courses on specific topics may also be available at universities, where students are generally required to be in residence for the duration of the course.

Experience is also related to training, in that provided adequate supervision is exercised the recruit may be expected to become increasingly skilled as he acquires more experience in carrying out the various tasks that he is given. In this respect, it is important that the supervisor should encourage rather than badger him to improve his performance. For the first few months, critical control must be applied with considerable discretion, little being gained by allowing the newcomer to execute a particular task many times if it is not correctly carried out. Corrections must be made as and where necessary so that the individual's work experience is enriched and he becomes a better workman in the process. As the training progresses, the degree of supervision exercised may be relaxed, but it is important that the supervisor shall still be seen to be interested in the recruit and his work.

With the more formal types of training, such as those carried out at colleges of further education, technical colleges and polytechnics, it is usual to link the education and training given at the college to the training and experience received at work. A number of the professional institutes concerned with the training of technicians and technologists now insist that candidates preparing for qualifying examinations should follow an approved and supervised curriculum at work conjointly with their theoretical studies and practical training at college. With many science-based courses it is impossible to allow sufficient time for the development of the practical skills that may make all the difference between a barely adequate and a very good laboratory worker. This is particularly so where practical technique needs to be combined with speed and accuracy. Careful supervision and time for much repetition, which the college may well be unable to provide, are often required. To these ends, many colleges have appointed course tutors or industrial liaison officers who spend a proportion of their time visiting employers' premises and meeting the students at work. In this way, the integration of training at work and education at the college may be greatly improved.

In the past, it was customary for employees who wished to study for either basic or improved qualifications to attend classes for a number of evenings per week, often over several years, until they were able to satisfy the examiners. This particular system has very largely been superseded in many countries by somewhat less arduous and, many would say, more efficient systems of study for professional qualifications. In conjunction with the major employers in their areas, many education authorities have now adopted schemes which utilize various forms of day or block release from normal employment for the student undergoing vocational training or education. Indeed in many areas it is now almost impossible to study certain subjects other than by means of a day or block release course.

With day release, the student is allowed one or more days'

absence from work per week with pay, so that he may take an approved course, during term time, at a local technical college. In addition to the day spent at college, he may also be expected to contribute an evening of his own time. Often the evening tuition falls on the same day of the week as does the day release from normal work. Although the student only has to attend at the college on one day per week, the instruction is spread over some 11 or 12 hours and many students may be expected to find this as much as they can absorb in such a period. However, this system certainly has compensating advantages. With block release, the student is allowed a similar total number of days off per year for study at college, but he generally takes them in one or two relatively long and continuous periods.

Both day and block release have their particular advantages and disadvantages for the employer and the employee. It often happens that where day release is the favoured method all the junior laboratory staff of a particular employer are likely to be away from work on the same day. This may result in considerable disruption being caused to the employer's business. With block release, it may be easier to arrange for each member of staff to be allocated to a different block, but much will depend on the total number of students taking the course at a given college. Block release provides for relatively long periods of continuous training or education, interspaced with uninterrupted periods at work. Although this may lessen the task of the laboratory manager when planning his staff dispositions, it may also be rather hard on the student, who will have to absorb a great deal of information during his fairly lengthy periods at college.

Most organizations are prepared to release their younger and junior employees to study for recognized qualifications but, understandably, they generally insist that the course taken must be directly related and relevant to the work, or necessary to enhance the employees' promotion prospects.

Sandwich courses are the reverse of block release courses. The student spends most of his time at college and is released for periods of training in industry. The proportion of time spent at college varies with different courses and this gives rise to the description 'thick' or 'thin' sandwich. In the UK, the sandwich course system is widely used for Higher National Diploma courses, particularly in engineering subjects where sometimes the student may be 'sponsored' by one of the larger industrial organizations by drawing a salary whilst on the course.

With all courses there is always the possibility that a particular student will fail to reach the necessary standard in the end-of-year assessment made by the college or, in an extreme case, to fail one or more of the externally set examinations. Because of the considerable investment made in time off from work, together with payment of college and examination fees, the personnel department will wish to examine the position rather carefully before necessarily allowing these costly arrangements to continue. Persons in positions of authority who have to make such evaluations will themselves have taken examinations at some time in their career and will be only too familiar with the realities of this type of situation. Although the majority of students expect to pass the examinations there may be many good and valid reasons for failure, not the least of these being sheer bad luck with the particular questions that were set. Whatever the apparent reasons for failure, it is important that there should be a thorough review of the student's work before he is allowed to continue with, or repeat part of, the course.

In carrying out this review, college end-of-term reports and direct contact with the course tutor or liaison officer may be found helpful. One must decide whether or not there was some justifiable or special reason for the failure; for example, had the student missed a number of classes due to illness; were the results for the whole class poor; was the examination unusually difficult; or was the student idle and therefore unworthy of further investment? Many organizations will happily pay all college and examination fees for a first attempt but only half the fees if the same examination has to be taken on a second and subsequent occasion. Others expect the student to pay all further fees if a second or further attempt is necessary.

There is one other disciplinary aspect of day release that needs to be considered and that concerns the question of regular attendance at classes. Colleges of further education usually have arrangements for speedily notifying employers of the absence or late attendance of day release students from particular classes or lectures. Figures for the percentage of possible attendance attained are also included in the annual reports made by the college to the employer. It is important that supervisors, or in large organizations the personnel officer, should be seen to maintain an interest in these reports and if necessary to take the appropriate disciplinary or supportive action. It is equally important that, except in the most extreme circumstances, neither pressure nor encouragement should be applied to students by departmental managers to cause them to absent themselves from lectures in order to cover sudden emergencies such as gaps in the labour force due to sickness or holidays among other staff.

Before taking disciplinary action against staff who are alleged to have arrived late or failed to attend a particular class, the supervisor should be sure that the report he has received is factually correct. Sometimes a student may attend a particular lecture and forget to sign the register and is therefore incorrectly reported as absent. In other circumstances, the lecturer may incorrectly call the roll or mistake one student for another with much the same result. Persistent lateness or failure to attend classes after due warning can only result in the withdrawal of day release facilities from the student in question.

MOTIVATION IN TECHNICAL EDUCATION

For many years now training schemes have existed for the benefit of technicians and others who wished to develop their skills and knowledge in the furtherance of their careers. An early example of such a scheme was the establishment of mechanics' institutes at the time of the Industrial Revolution for the purposes of giving instruction, mainly in the engineering trades, to apprentices. In more recent times, there has been a proliferation of colleges of further education, technical colleges, polytechnics and institutes, etc., where students are able to follow a very wide range of courses leading to different qualifying examinations.

Some of these schemes have worked extremely well, bringing great benefit both to the student and to the community at large, whilst others have been less successful. One of the difficulties often encountered in part-time education has been the lack of 'open end' examinations and courses. Frequently a student would find that having successfully passed a particular examination, no natural and logical progression existed to take him to the next stage. Sometimes it was necessary to repeat much of what he had already studied in order to comply with the syllabus of a further examination. This position was frequently exacerbated by the rigid entry requirements of many examinations. Occasionally, various entry requirements were waived or exemptions allowed, but the system has undoubtedly led to much wastage and frustration. This has become particularly evident in instances where a student wished to change the direction of his career in the light of experience acquired at work.

It is useful to consider the motives of students who set out to take qualifying examinations by means of part-time study at technical colleges. For some, the intellectual stimulus will be sufficient in itself, whilst others will be seeking ways of improving their job performance and career prospects. The majority will be motivated by a complex mixture of these and other factors which will almost certainly change in the course of time or to which they will attach different levels of importance as their career develops.

At the beginning of their career, many people find work to be its own reward, whilst others are driven on by the prospect of a better job or increased earning power. Some never lose the feeling of satisfaction in doing a job well; others come to consider that a better job will provide the wherewithal to support increased family responsibilities and expanded leisure activities.

To these ends, managers and staff should consider, from their respective viewpoints, the inducements and rewards that may be given or received following a successful course of study or training. In certain industries or organizations there is a direct financial reward attached to the passing of specific examinations; in others, promotion is barred without them, and in some the passing of the appropriate examinations is regarded merely as qualifying the person for promotion. He then must wait for a suitable vacancy to occur.

RECENT DEVELOPMENTS IN TECHNICAL TRAINING AND EDUCATION

To overcome some of the difficulties previously mentioned there has been much current rethinking of what is required by way of technical training and how it may best be provided in the present economic circumstances. The advent of the Open University with its modular system of study and its ability to tailor degree courses to the particular needs of individuals, together with its willingness to consider a wide range of entry qualifications, was a great step forward, although many will consider that its full impact has yet to be felt.

The Technician Education Council (TEC) was established in 1973 in fulfilment of the recommendations of the Haselgrave Committee which reported in 1969. It is intended that TEC's function will be essentially to rationalize the existing provisions for technician training thus saving valuable resources and, further, to provide a system of technician education which is responsive to both industrial requirements and to students' needs. TEC will award certificates, higher certificates, diplomas and higher diplomas. It is intended that the existing City and Guilds Part I Certificates in the various scientific fields and the Ordinary Certificate in science subjects will be replaced by a Technicians' Certificate in Science, whilst the City and Guilds Parts II and III, together with the Higher National Certificate, will be replaced by the Higher Technician Certificate. In a statement of policy issued in 1974[5], the Council has said that 'its diplomas will be extensions of its certificates in the sense that while the depth of technical content will often be similar, diploma programmes will be broader than those leading to certificates. A system of "units" is being developed and it is expected that certificate courses will commence in 1977. The diploma courses will follow at a later date'. From study of the programmes that have so far been announced, it would appear that in future the student will have a far better chance to pursue a course that will not only satisfy his immediate needs but one that may be adjusted, without waste, as his career develops.

There has been much vocal opposition by individuals and organizations interested in maintaining the present system of qualifying examinations, just as there has been in the past when almost any other new system has been introduced. Much of this opposition has been subdued in the normal process of consultation but it remains to be seen, however, just how useful and effective the new system of technical education will become.

The Technician Education Council will operate in England, Wales and Northern Ireland. For Scotland, the comparable responsibilities for technical education will be exercised by the Scottish Technician Education Council (SCOTEC) which was also established in 1973. Although the objectives and standards of the two organizations are expected to be broadly similar, there are undoubtedly differences in the timing and standard of primary and secondary education that will have to be taken into account when setting up arrangements for technical education in Scotland. Both councils have said that it is their aim to seek compatibility of course arrangements and standards for their awards and to keep such matters under review. In addition, it is envisaged that there will be collaboration over the planning and operation of schemes which are highly specialized and involve only small numbers of students.

Laboratory discipline

Whatever one's personal attitude, it now has to be accepted that the authoritarian approach to discipline at work is no longer acceptable. This is particularly so in laboratory life where the authoritarian system would seem to be not only inappropriate but counter-productive. To be effective, discipline may no longer be imposed from above without regard for the views and feelings of those on whom it is being imposed. Discipline brought about by leadership and example on the part of the management team, is far more likely to be effective and productive than was the previous system, which relied on threats of dismissal and the application of other punitive measures. Discipline today must be seen to be a product of the application of common sense and reasonable behaviour, i.e. a means of producing order in the general interest of employer and employees alike. Furthermore, it has to be seen to be fair and must be generally acceptable to the work-force. No longer is it acceptable to have one rule for the senior staff and another for the juniors. Clearly a manager who habitually arrives late for work, without compensating in other ways for the time apparently lost, will find himself in some difficulty if he has to reprimand a junior member of staff who is unpunctual.

Leadership and example are two of the most important factors that must be deployed in bringing about a system that is seen to be just, and in the interest of all, but is is possible to carry this process too far. We are all familiar with the laboratory where it is possible to distinguish the senior staff from their fellows by their willingness to work longer hours or to perform difficult or unpleasant tasks when others are unavailable or disinterested.

Discipline can be related to the style of the manager. On the one hand we have the manager who tells, sells and consults, leaving a large area of freedom for his subordinates to think and act for themselves, whilst on the other hand we have the authoritarian manager who delegates nothing to his staff but expects them to do exactly what they are told to do and nothing else. Not only is he constantly 'breathing down their necks' but everything has to be referred to him for decision. Somehow a happy medium must be found where all staff know what is expected of them and how far they are to be left to do their best and get on with the job. Lack of effective discipline can only lead to lowered morale, falling output or production and ultimately to the destruction of the organization[6].

Reprimands can vary from a mild cautionary word to the severe 'telling-off'. The object in giving them is to correct the performance or behaviour of the individual concerned. For anything other than the most minor offence, they should be given in private and never in the heat of the moment. A display of temper on the part of a manager will only antagonize the work-force. In the uniformed services, the penalty of a reprimand to an individual is a means of placing on record that he has committed an offence and such an occurrence may severely limit his future chance of promotion. In civilian life, however, reprimands are usually given in private to inform an individual that he has departed from the accepted code of conduct expected of him.

Formal disciplinary interviews are held (a) to establish the facts, (b) to give the person concerned the chance to explain or refute the allegations made against him, and (c) to ensure that the offender is aware of the seriousness of his transgression. The prime objective is to ensure that the same or a similar offence is not repeated. Many organizations have, in consultation with staff representatives, negotiated an agreed code of practice that specifies the procedure to be adopted when disciplinary action is to be taken against a member of staff. It is imperative that all action taken by management under such an agreement shall conform exactly with the procedures laid down. For a major offence, it is normal to allow the employee to have his (Trade Union) representative with him[7], but for less serious transgressions, this procedure may sometimes be dispensed with in favour of a face-to-face discussion where both sides may be less inclined to adopt rigid attitudes.

For minor departures from the rules, it is normal to give verbal warnings, but for repeated minor offences or for the more serious offence the offender is usually warned in writing by a senior member of staff that, if his conduct or performance does not improve within a given time, he will render himself liable to suspension or dismissal. It is important that once such a warning has been given, management should be seen to take an interest in the person's subsequent behaviour if only to ensure that he is given every encouragement to reform.

In the UK, the employee's rights are covered by the Employment Protection Act 1975[7], the Trade Union and Labour Relations Acts 1974 and 1976[8] and the Contracts of Employment Act 1972[1]. Under the Employment Protection Act and the Trade Union and Labour Relations Act, an employee with more than 26 weeks' continuous service is entitled to obtain from his employer a written statement of the reasons for his dismissal. The employee can complain to an industrial tribunal if the employer refuses to supply a statement or gives reasons which are inadequate or appear to be untrue. He can also complain to the tribunal that he has been unfairly dismissed. If his complaint is upheld, the tribunal may order his reinstatement in the same or a comparable job. If this is not possible, the tribunal will consider the award of compensation for 'unfair dismissal'. Under the Contracts of Employment Act, most employees are given the right to a minimum period of notice of termination of employment, according to length of service. They are also entitled to be given a written statement of their terms and conditions of employment covering rates of pay, holidays, sick leave and pension entitlement, etc.

Having in mind the four Acts of Parliament mentioned

above and the need for agreed and published disciplinary procedures, it is important that the agreed procedure must be complied with to the letter when a member of staff is to be dismissed. There must be sound and compelling reasons for dismissal, and adequate records must be kept of all relevant events leading up to the dismissal. It is unlikely that vague reasons such as 'insolence', 'laziness' or 'misconduct' will satisfy an industrial tribunal that a person has not been 'unfairly dismissed'.

In the past, members of staff whose performance was considered to be unsatisfactory have sometimes been persuaded to leave by warning them that they were unlikely to obtain promotion or a better job in the foreseeable future. Occasionally the pressure to leave may have been increased by the application of other rather more dubious practices such as giving them the most unpleasant tasks to perform. Not only is this practice to be deplored (if only because it is bad management) but it must now be discontinued because it is illegal. Although the person resigns apparently of his own free will, there is always the possibility that at a later date he could justifiably complain to an industrial tribunal that the pressure put upon him to resign was unfair and amounted to 'constructive dismissal'. It is still possible for an employer to give an employee who is guilty of gross misconduct or of a major breach of discipline, the option to resign rather than be sacked. However, the evidence of misconduct must be substantial and the employer must recognize that he may have to justify his action before an industrial tribunal.

Termination of employment

It is important that records are kept of all persons who leave the organization. When a person resigns, he should be interviewed by a senior member of the management team who should make a determined effort to find out the true (which may not always be the same as the stated) reasons for leaving. If one department has a higher turnover of staff than the others, then management must discover the reasons for the discrepancy. The work may be uninteresting, unrewarding or particularly difficult. Sometimes it may be that the local manager or supervisor is at fault in that the training or supervision is poor. Occasionally it will be discovered that there is a trouble-maker at work. Whatever the reasons or causes for the high turnover of staff, they must be ascertained and, if possible, corrected before a general malaise develops. In general, a high labour turnover indicates low morale.

Before a person leaves it must be ensured that all keys and notebooks, etc., are handed over to a responsible person and that books on loan from the library are returned. On the final day, any work in progress must be left in a safe condition and proper arrangements made for its eventual continuation. The work bench must be cleared and all samples, etc., properly labelled, safely stored and indexed. In the weeks prior to his departure, instructions for the use of any special equipment or laboratory methods that he has been using will have been compiled so that there may be a smooth takeover of the work by the replacement member of staff. Where possible, there should be an overlap between the departure of the old and the arrival of the new employee. This is sometimes difficult to arrange and is not always convenient, in which case it is essential that all important information is given to an intermediary. It is assumed that the personnel or accounts department will have dealt with the final salary cheque, holiday pay, income tax, National Insurance and pension arrangements and will have obtained a forwarding address.

After a person has left his employment, many organizations require their managers to complete a questionnaire concerning the standard of service he has given. This will provide a permanent record which can be referred to at any future date should references be requested. Sometimes the most significant question on this form is: 'If this person were to apply for another post in the company, would you employ him again?'

REFERENCES

1. The Contracts of Employment Act 1972. HMSO, London
2. Scott, B. and Edwards, B., *Appraisal and Appraisal Interviewing.* The Industrial Society, London (1972)
3. The Race Relations Act 1968. HMSO, London
4. The Sex Discrimination Act 1975. HMSO, London
5. The Technician Education Council, *Policy Statement.* TEC, London (1974)
6. Foxen, T., *Effective Discipline.* The Industrial Society, London (1977)

7. The Employment Protection Act 1975. HMSO, London
8. The Trade Union and Labour Relations Acts 1974 and 1976. HMSO, London

FURTHER READING

Betts, P. W., *Supervisory Studies,* 2nd edn. Macdonald & Evans, London (1973)
Cooke, A. J. D., *A Guide to Laboratory Law.* Butterworths, London (1976)
Foxen, T. and Smith, E., *The Manager as a Leader.* The Industrial Society, London (1976)
Fraser, J. M., *Employment Interviewing.* Macdonald & Evans, London (1966)
Gode, W., *Training and Staff.* The Industrial Society, London (1972)
Henderson, J., *A Guide to the Employment Protection Act.* The Industrial Society, London (1975)
ISTOX Laboratory Management Symposia: *I. The Role of the Chief Technician and Training for Management* (1977); *II. Staff Recruitment and Selection* (1978). The Institute of Science & Technology, London
Lawrence, K. C. *Personnel Management.* Hutchinson Educational, London (1972)

3
Purchasing and Financial Control

Sources of income

The precise arrangements made to provide the funds necessary to staff, equip and run a laboratory depend very largely on the type of institution or organization with which it is associated and on the principal source of income of that body. For convenience, four of the most commonly employed arrangements are examined here:

(1) Governmental laboratories, including state hospitals and school laboratories, are provided with core funds for salaries, equipment and materials on an annual basis. The budget is reviewed at the start of each financial year and other funds may be made available for specific projects. A finance committee is responsible for preparing the budgets and an accountant handles the details such as payment for goods and services. The laboratory manager is involved in presenting estimates for his department or section and is responsible for controlling expenditure.

(2) Industrial laboratories are allocated a sum of money each year for general expenditure and further amounts for specific projects, such as the development of a new product. Project funds are usually provided for a specific project related to a definite period of time after which the company will decide whether further expenditure is justifiable or not, depending on the state of development reached and on the return expected from the new product.

(3) Research establishments such as universities, medical schools and scientific societies receive much of their finance in the form of grants from government sources, fund-raising organizations and charitable trusts. These grants are usually made to an individual or group working on specific problems and money is allocated for clearly defined purposes such as salaries, capital equipment and running costs. Many university departments base their finances on a quinquennium, i.e. a five-year period, at the start of which funds are allocated for the whole period. Adjustments of a comparatively minor nature are made at the start of each financial year. Although this system has some advantages, it does require considerable expertise in terms of forward planning.
(4) Contract work laboratories carry out projects for other organizations which lack the necessary facilities or expertise to do the work themselves. This work usually takes the form of research into products or materials, analytical control or a diagnostic service. The client is charged on a cost plus overheads basis or at an agreed fixed price.

Some laboratories may receive funds from more than one source, such as from fund-raising organizations, trusts and industrial organizations. Regardless of the source of income, it is necessary to apply for funds to be allocated at the start of each financial year or prior to the commencement of a specific project. The usual practice is to determine as accurately as possible the sums needed for each purpose and for this one must ascertain the number and grades of staff involved, the length of time for which the project will run and the equipment and materials needed. If a period of more than one year is being considered, allowances are made for salary increases and inflation which may represent a significant sum.

Grant-giving bodies generally demand that applications are made on a form which they provide. Specific details of the work envisaged, accurate costings of the equipment and materials needed, together with salaries and other expenses to be met must all be enumerated. Assuming the application is viewed favourably, the grant giver will, in due course, make funds available either to the applicant or to the employing authority and will probably state the method to be used in accounting and making payments. Sums may be allocated annually or at other intervals and a statement of expenditure may be required at these intervals. The grant holder is also expected to provide progress reports during the period of the grant.

In cases where the laboratory is part of some larger organization, the accounts department will have the main responsibility for handling finance, particularly salaries, but the laboratory manager still has the task of ensuring that funds are correctly spent and that records are kept of all transactions. The techniques for doing this vary in detail but the basic principles are fairly straightforward and will be discussed later.

As expenditure is a continuing process, it is important to maintain a system which continually monitors the state of the finances. One can then ascertain rapidly the amount of money which has been spent to date and the sum available for future use. Running out of funds can bring work to a standstill and money unspent at the end of the financial year may be lost. A simple way to exercise control is to produce a graphical representation of actual spending rate and to compare this with the theoretical rate (*Figure 3.1*).

As the year progresses, one can read off not only the amounts spent and still available but the spending rate can be ascertained from the graph. It is then possible to advise those concerned that they are in danger of running out or leaving funds unspent by the end of the year. It is usually found that expenditure rate is high in the early stages, when equipment and materials are purchased, and tends to slow up later.

Expenditure recording in its simplest form consists of a written record of all purchases and expenditures together with the minimum of relevant details, i.e. the order number, date of passing invoices and the name of the supplier would be sufficient. Each grant or budget is recorded in its own ledger or its own section of a ledger. Any tax or duty which may be recoverable is entered separately.

It is important for the laboratory manager to maintain a good working relationship with the accounts department as there are frequent complications in transaction with suppliers, such as incorrect deliveries, overlooked invoices

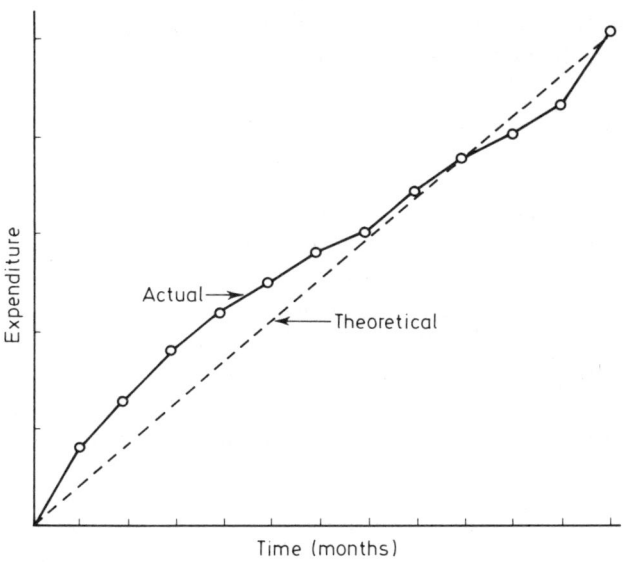

Figure 3.1. Control of financial expenditure. The actual cumulative expenditure is shown plotted as a solid line and may be compared with the theoretical or budgeted sum (dashed line). At the end of the financial year the two lines should coincide

and so on which cannot possibly be resolved unless there is good communication in both directions.

Purchasing

Before any purchase can take place, two things must be established; first, the need for the items in question and, secondly, the availability of funds. In a large organization, requests are frequently made for items which, unknown to the person making the request, are either already in stock or are available on loan from another department which is not using them.

Assuming the need does exist and money is available, decisions have to be made regarding which equipment is the most suitable to buy and from which supplier it should be obtained. Laboratory supplies can be divided into capital equipment, that is to say major equipment from which several years of service can be expected and which can be covered by insurance, and consumable materials, such as chemicals, glassware and disposable items. The factors affecting the choice of either are as follows:

CAPITAL EQUIPMENT

(1) Price, including extras such as delivery and installation.
(2) Suitability for the work envisaged.
(3) Quality of workmanship.
(4) Reliability under working conditions.
(5) Convenience in use, including physical size, adaptability for future use, supplies needed for operation, etc.
(6) Delivery.
(7) After-sales service and availability of spare parts and accessories.
(8) Safety in design and construction.

When buying expensive equipment, it is advisable to contact several companies and arrange for discussion with representatives, who will quote a price for their equipment. In most cases demonstrations can be arranged. Some companies are prepared to supply the addresses of other organizations which have bought their equipment and which would be prepared to comment on its performance.

Prices can sometimes be misleading, especially if taken from catalogues; increases have often been applied before the catalogue is published. In some cases, extra charges are levied for delivery, particularly when buying from overseas. Import duties and other taxes may be applied and installation is sometimes charged separately. Quality of workmanship is not always apparent, particularly with encased instruments, but the supplier should be prepared to allow a thorough examination and provide a meaningful guarantee of free repair or replacement. Under UK law, it is illegal for a supplier or manufacturer to make misleading statements about his product or to supply an item not suited for the purpose for which it is sold[1-3].

Some equipment may be thoroughly reliable when used in the laboratory but if required for field work, for example, may not survive the adverse conditions encountered. A domestic washing machine acquired for use in an animal house had, after one year, the equivalent of 10 years' domestic use. It would obviously have been far more satisfactory to have bought a commercial machine in the first place. Points such as this should be established before purchasing. Equipment required for use in a particular room must, of course, be of such a size as to enable it to pass through the doorway and of such weight as the floor will support safely. Should it require power or water supplies, the availability, or cost of adding these, must be determined.

Delivery of an item may be from stock or there may be a delay of several months from one supplier compared with a few weeks from another — there will be instances where a long delay is unacceptable and other instances where this is of no importance. Most apparatus requires servicing at some stage and if the supplier or manufacturer is unable to provide this service, an expensive item can become worthless. If the equipment is manufactured overseas, an agent with servicing expertise may or may not be available or may not be as satisfactory as the service which one would expect from the manufacturer.

The design of laboratory equipment usually ensures that it does not present a safety hazard, but there have been examples of apparatus which is either badly designed or poorly assembled and is dangerous. In the UK, the Health and Safety at Work etc. Act 1974 makes it illegal to supply unsafe equipment (see Chapter 7).

CONSUMABLE MATERIALS

(1) Price, including extras and discounts.
(2) Quality and repeatability.
(3) Delivery.
(4) Size of standard packages.
(5) Collection of empties.

When buying small amounts of consumables the price is not of great significance, but when buying in bulk or negotiating long-term contracts, it becomes a very important factor. Most suppliers offer substantial quantity discounts which increase with the size of the order. It is advisable to calculate the actual 'each' price of the items as the phrases used in some price lists can be misleading. One has to remember also that bulk buying must mean bulk storage, and facilities for this may be costly or may not be available. Some materials are also prone to deterioration in storage and the end results of an apparent saving may be a loss.

The price of consumables is often reflected in their quality, and a cheap item may be inferior. This applies particularly to chemicals, the purity of which may render a cheaper product useless for a particular application. Glassware may be used in conjunction with a piece of apparatus which it is designed to fit and, if the design is changed, the apparatus may have to be modified or replaced. If materials are required quickly it will be necessary to buy from an organization which can guarantee rapid delivery. There are numerous items which have to be obtained from foreign manufacturers and if the local agent is out of stock delays can be considerable. Certain manufacturers will only supply their products in standard packs, so if only small amounts are required it will be necessary to buy from a distributor who is prepared to sell smaller quantities. Glassware is usually sold in standard pack cartons which also form a convenient storage pack (see Chapter 4).

ORDERING PROCEDURE

The cost of processing orders is considerable and many suppliers find it necessary to impose a minimum order value on their products to cover this cost. This can be a problem for the buyer when only a small quantity of one item is required, and it is possible for 25 g of a particular chemical to cost the same as 1 kg. It is sometimes feasible to delay ordering until other items are required.

Information on sources of supply is fairly widely available and includes manufacturers' catalogues and advertising literature, professional journals, published papers, exhibitions and the buyers' guides distributed in many countries. Trade names are so numerous as to be impossible to remember and a card index of these is easily established and extremely useful. Items of low value which are easily obtained locally do not warrant the expense of raising purchasing orders and the documents which result; such items are usually purchased for cash and paid from the petty cash account. Internal transactions between one department and another or between a laboratory and stores require a written requisition, signed by an authorized person. This enables accounts to be kept in order and ensures that the transaction is recorded should a query arise at a later date.

The frequency with which purchasing orders should be sent out depends on the size of the organization and on the nature of the work. Where requirements are predictable, such as in a routine department, it may only be necessary to order once a month but in the case of a large research organization, where the results of each day's work may determine the requirements of the following day, it is usual to send out orders daily.

Purchasing from an outside source requires a number of documents, each of which serves a separate purpose. Initially, a request is made by the intended user to the laboratory manager or other person responsible for buying. This request should be in writing and signed. The buyer then translates this to an official order which carries an order number, and this will be used as a reference on all subsequent documentation relating to the order. The order is sent to the supplier and one or more copies are retained. The order must state the number of items required, an unambiguous description and the catalogue number, together with instructions regarding delivery, invoice address and any other relevant information. If goods are required urgently, suppliers will often accept telephone orders, asking for the written order to be sent on later, marked 'confirmation of telephone order, do not duplicate'.

If delivery is likely to be delayed, the supplier will probably send an acknowledgment or confirmation of order, stating the anticipated delivery date. When the goods arrive, they are accompanied by a delivery note, listing the items and, in the event of the delivery being incomplete, a list of items to follow. The goods are checked against the delivery note and the copy of the original order marked and preferably dated to indicate their receipt. The supplier must be notified of any shortages or damage. The delivery note is kept for future reference. It is a useful practice to mark all copy orders in some readily obvious manner when delivery is complete, such as removing one corner of the document. This facilitates retrieval of incomplete orders when they need to be 'chased'.

Following delivery, the supplier will send his invoice which should be checked and, if correct, passed for payment with information regarding which budget or grant is to be debited. At the same time this information is recorded in the ledger. Any errors should be notified to the supplier without delay; if goods have been returned or overcharged the supplier will furnish a credit note, which is also passed to the accounting section, and the sum involved credited to the appropriate budget.

At regular intervals, usually monthly, the supplier will send a statement, which is a resume of all transactions and includes all invoices, credits and outstanding accounts which have accumulated during the period. The terms of business of suppliers vary in detail but most require settlement within 30 days, and it is normal business practice for the accounts section to issue cheques on a given day during the month and send these to the suppliers.

From time to time, queries arise due to incorrect deliveries, invoice errors, goods damaged in transit, mislaid documents and so on; in such circumstances, the order number is the vital link between buyer, supplier and accountant. Details of any correspondence or telephone calls should be noted on the copy order, a system which can save a great deal of time when a query arises, perhaps several months after the original order was placed.

The law regarding purchasing is fairly complex, particularly when an overseas supplier is involved. In the UK any purchase is regarded as a contract, whether an order is placed in writing or verbally, and as such is covered by a variety of legislation with which the buyer must familiarize himself. If for any reason an order, once placed, has to be cancelled this must be done in writing.

SPECIAL PURCHASING CONSIDERATIONS

In most cases purchases are perfectly straightforward and require little more than accurate recording, but there are instances where other factors have to be taken into account.

Ethyl alcohol

In the UK, ethyl alcohol is subject to duty applied by Customs and Excise. Where this material is required in the laboratory as a reagent or solvent, it is possible to obtain licences exempting the purchaser from this duty. Application is made to the local office of Customs and Excise stating the name and address of the laboratory, the purposes for which the alcohol is required and the expected annual requirement. If exemption is granted, it is essential to comply with the regulations regarding receipt, storage and issue, all of which must be recorded. A locked store is necessary and duty-free ethyl alcohol must be stored separately from other materials. For this purpose a locked cupboard within the store is usually acceptable. Records must be available for examination by a Customs and Excise Inspector and any infringement of the regulations may result in legal action and a withdrawal of the licence.

Recovery or redistillation of ethyl alcohol is not permitted, except for the immediate purposes of the work being done.

Industrial methylated spirit does not attract duty but Customs and Excise permission must be obtained before it may be purchased. Its issue must be recorded as with ethyl alcohol.

Poisons

A considerable number of substances in common use in the laboratory are legally defined as poisons and in the UK are subject to one or more Acts of Parliament[4-6]. In general, they may be purchased from authorized suppliers against an official order from a recognized establishment, but some chemicals which are also used for medical purposes are subject to more complex regulations[7-9] and their receipt, issue and use must be recorded. For some the purchase order must be signed by a registered medical, veterinary or dental practitioner. Most chemical suppliers and manufacturers indicate in their catalogues which of their products are subject to the various regulations and are prepared to advise on problems which may arise in relation to their purchase.

Importation

Although most laboratory supplies can be obtained from home sources, there are occasions when it is necessary or advantageous to purchase from an overseas manufacturer. Where possible this should be done through an agent who will handle all the documentation and, in the case of major equipment, will usually provide maintenance or servicing facilities. If no agent exists, the purchaser will need to prepare the necessary documents and satisfy himself that alternative arrangements can be made for servicing in the case of breakdown or malfunction.

In the UK, import duty is applied to materials and apparatus purchased from most foreign sources but in certain circumstances duty-free importation may be granted to non-profit-making organizations[10]. Application is made to the Department of Trade and Industry on form DFA3, stating the purpose for which the goods are required (normally research or education), which alternatives of UK manufacture have been considered and the reasons why the foreign product is preferred. If satisfactory answers are given, such as would be the case if no UK equivalent existed or if the foreign model were known to be superior in performance, duty remission may be granted subject to several conditions. It is neither permitted to resell the items nor to use them for any purpose not specified in the application. Under certain circumstances, it is possible to pay the import duty and to reclaim it at a later date.

Purchase of animals

Many biomedical research and control laboratories use animals and whilst most of these are bred by domestic specialists, some are imported either directly by the purchaser or through an agent. Recent legislation in the UK[11] is concerned with the conservation of endangered species and prohibits their importation without the appropriate licences. At the present time the UK is free of rabies, and further legislation places restrictions on the importation, movement and housing of all mammals as these could possibly be carriers of the disease. Such animals may only be accommodated in approved and licensed quarantine premises, separated from non-quarantined animals, a separate authorization being needed for each movement. Application for the prescribed documents must be made to the Ministry of Agriculture, Fisheries and Food before these animals are ordered.

No animal, whether or not subject to the Rabies Act, may be purchased for research purposes unless the premises have received prior approval and registration under the Cruelty to Animals Act 1876 (see Chapter 6).

Radioactive materials

The purchase, use and disposal of radioactive substances is controlled by the Radioactive Substances Act 1960[12] (see Chapter 6) and before such materials can be purchased it is necessary for the premises to be registered. Registration is only granted to laboratories with proper facilities for radioactive work and storage and with facilities for the safe disposal of radioactive waste. Limitations are imposed on the substances which may be purchased, on the amounts which may be held at any one time and on the amounts which may be disposed of by approved methods. All consignments received and issued to workers must be recorded and the records must be kept available for inspection.

COST CUTTING

Whatever the source of the laboratory's finance, it is one of the manager's most important functions to economize whenever possible. Money saved in one direction can be used to advantage in another. There are many ways in which this can be achieved without prejudice to the quality or quantity of the work done. Obviously, intelligent buying has a prominent part to play. Bulk buying has already been mentioned and the savings brought about through quantity discounts can be very considerable, provided storage space is available (see Chapter 4). It is frequently possible to negotiate with suppliers for special discounts, particularly when buying for an educational or other non-profit-making organization.

Buying cheap apparatus and materials can prove to be more costly than the apparently expensive equivalents, as the quality is usually inferior and replacements must be made more frequently. Materials which are required throughout the year can often be the subject of a contract purchase, in which case consignments are delivered at regular intervals, a discount being allowed for the whole amount bought during the year.

Pilfering and theft can account for the loss of considerable sums of money and steps can be taken to reduce this to a minimum. Observation of the work can often reveal areas of wastage, e.g. excessive glassware breakages should be investigated and careful use of equipment must be encouraged to reduce costs of repair and replacement. Techniques using costly reagents can sometimes be scaled down thereby using less material. Disposable plastic ware can in some instances replace glassware, thus avoiding the cost of washing and/or sterilizing. Salaries are almost invariably the most expensive item on any budget and staff not fully occupied should have their work schedule reorganized. This can often be advantageous to the employee as it introduces further variety into his working day and, in the case of a young person, may assist with his training.

It is important to avoid petty cost cutting to the point where it becomes an obsession. Not only does this become a constant source of annoyance to others but it is in concentrating on such minute details that one remains unaware of the greater problems.

REFERENCES

1. The Sale of Goods Act 1893. HMSO, London
2. Supply of Goods (Implied Terms) Act 1973. HMSO, London
3. Consumer Protection Act 1971. HMSO, London
4. The Poisons Act 1972. HMSO, London
5. The Poisons Rules 1972. HMSO, London
6. The Poisons List Order 1972 and 1974. HMSO, London
7. The Pharmacy and Poisons Act 1933. HMSO, London
8. The Dangerous Drugs Act 1965. HMSO, London
9. The Food & Drugs Act 1955. HMSO, London
10. The Import Duties Act 1958. HMSO, London
11. The Endangered Species (Import and Export) Act 1977. HMSO, London
12. The Radioactive Substances Act 1960. HMSO, London

FURTHER READING

Guy, K., *Laboratory Organisation and Administration*, 2nd edn. Butterworths, London (1973)
The Trade Descriptions Act 1968. HMSO, London
Value Added Tax (Donated Medical Equipment) Order 1974. HMSO, London
Value Added Tax, HM Customs and Excise Notice No. 712. HMSO, London

4
Management of Stores

The size of the store and the quantity and variety of materials carried must, of course, be related to the size of the organization it is to serve and the type of work undertaken. For the small laboratory, a single room may suffice, but the store required to supply a number of large laboratories in a group such as a university or hospital complex may reach the capacity of a substantial warehouse.

Regardless of its size, the principal functions of a store remain the same and may be summarized as follows: (1) to receive incoming goods; (2) to store goods in a safe and satisfactory manner; (3) to issue goods as required; (4) to record all transactions.

Stores policy

As all incoming goods are the result of buying transactions, it follows that good liaison between the buyer and stores manager is essential to the successful running of the stores. In many organizations both functions are carried out by the same person. It is customary to establish some form of buying, storage and issuing policy in collaboration with all concerned, namely, the buyer, stores manager, finance officer and representatives of the consumers. The latter are normally the senior technical staff of each department using the store. It is then possible to decide what goods are to be stocked and in what quantity, and it will be necessary to update the policy as conditions dictate. Changes will always be necessary because of such factors as alterations in the type of work being done, availability of materials, financial restrictions and so on.

In a fairly small organization in which the nature of the work is largely repetitious, it should not be difficult to forecast with considerable accuracy the requirements for future work and to maintain stocks accordingly. Conversely, in a large and complex organization engaged in a wide variety of work, the system must be under constant review. Research departments present a major problem in this respect as their requirements change almost daily and as a general rule require a large range of items in fairly small quantities. Teaching departments and laboratories engaged in routine work usually require a smaller range of materials but in larger quantities.

At no time should essential items be allowed to run out but stocks must not be so high that they are likely to deteriorate or to become obsolete before they are required, and a considerable amount of judgement based on experience of the particular circumstances is required in order to establish a satisfactory compromise between over- and under-stocking. Stock in hand represents a considerable capital investment which, if excessive, may result in a shortage of finance for other purposes; conversely, such an investment may appreciate in value in a relatively short time and act as a buffer against inflation.

Whilst the needs of the small laboratory can be supplied by one store, it is generally more efficient for the larger establishments to incorporate a system of stores, each with its separate function. Bulk deliveries are made to a central store from where they are issued in standard units to one or more secondary stores for distribution in some instances through further departmental stores. In many respects this is similar to the way in which the retail trade operates in distributing goods to the buying public (*Figure 4.1*).

For example, in a typical university or college, incoming goods are received by the central store and held until required by the various secondary stores, usually one for each faculty. From there they are issued to the department stores, from which they are drawn by the laboratory staff as required. At first sight this may appear to be unnecessarily complicated but such a system has a number of advantages. It simplifies the buying and receipt procedures as these are carried out by the stores staff and are, therefore, more easily controlled than if handled by numerous individuals. A chain of responsibility is established which reduces the possibility of errors and simplifies enquiries if errors should occur. The maximum possible benefits of bulk buying are obtained, because very large purchases can be made and space is provided for storage. Stock control is simplified as the bulk stock is held in one place rather than being distributed over a wide area. Laboratory staff are relieved of the problem of maintaining stocks and finding suitable storage space.

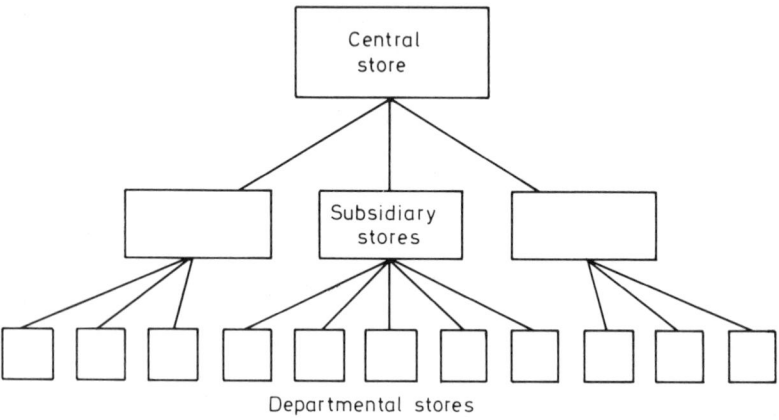

Figure 4.1. Distribution of supplies from central store to subsidiary stores and finally to departmental stores

The only real disadvantage lies in the delays which may occur in obtaining goods if the whole system is not managed efficiently. The precise details of the way in which the system is operated must, of course, be adjusted to suit the particular circumstances, and hard-and-fast rules cannot therefore be laid down. For example, in some situations it may be preferable for the stores staff to deliver goods to the departments and in others it may be more satisfactory for the laboratory staff to collect the goods from the stores themselves. Some form of compromise will frequently be found to be the most efficient answer to the many problems of stores management, but it is important to remember that the stores are there to serve the laboratories and the system must be sufficiently flexible to achieve this objective.

Stores staff must be selected with as much care as would be exercised when choosing laboratory staff. There is little doubt that a good storekeeper can exert considerable influence on the smooth running of the laboratories. Incorrect deliveries, unnecessary delays, chaotic documentation and low stock levels have a demoralizing effect on the whole establishment, resulting inevitably in a lowering of work output. The competent storekeeper will make every effort to ensure that at least those items in frequent demand are kept in stock and that a range of goods requested only occasionally are made readily available. In many cases, he will be able to make an intelligent substitute for items not readily available and in order to do this he must have a clear idea of the function of the apparatus and materials which he issues. He should also be aware of the hazards associated with various materials and of the special storage requirements necessary for stock which is likely to deteriorate. On numerous occasions he has to deal with people who are not at all certain of their needs and who have difficulty in conveying their precise requirements.

A large store will require one or more assistants of varying grades, at least one of whom should be sufficiently skilled to take over supervision in the absence of the storekeeper. A considerable amount of heavy work is involved and staff will be required for unloading and distribution. In many instances, the store will be expected to carry certain goods for use in areas other than the laboratories, e.g. spares and materials for the maintenance staff, stationery and office supplies, cleaning materials, etc., and the stores staff will need to become familiar with these requirements.

Stores design and planning

The basic design requirements are much the same for any store regardless of its size, and to avoid duplication the design of a medium-sized store will be considered (*Figure 4.2*). Larger or smaller stores are simply scaled down or up as appropriate.

The location of the store is of extreme importance and should be considered at the design stage (see Chapter 1). Because it serves the entire organization, it is a vital part of the building and as such it should receive at least as much thought as the laboratories. Goods are received frequently and in large quantities and the store should therefore be

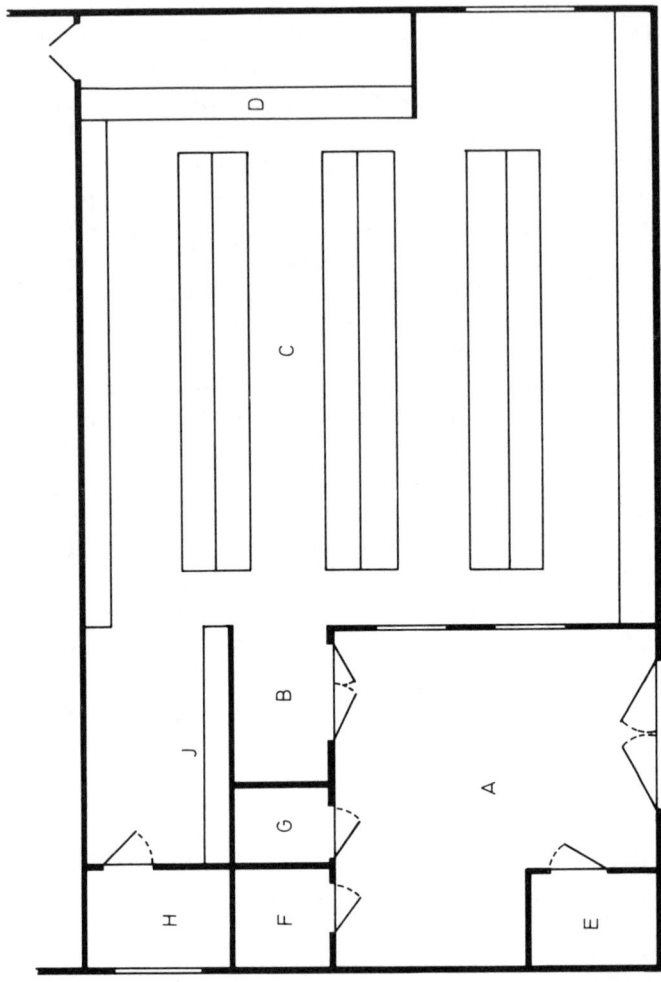

Figure 4.2. Suggested plan for an efficient laboratory store: A, unloading yard; B, unpacking and sorting area; C, storage area; D, issuing counter; E, solvent store; F, gas cylinder store; G, corrosives store; H, office; J, special storage area

situated on the ground floor, preferably at the rear of the building and with unrestricted access for delivery vehicles. A covered unloading area is a great asset because it protects both goods and staff from the weather — an important consideration as wet goods will deteriorate rapidly and wet containers are slippery and therefore constitute a hazard.

The storage area must, of course, be of adequate size[1] and the floor construction must be able to support the very considerable weight of goods involved. Good ventilation and heating are important not only to provide a healthy working atmosphere but also to minimize deterioration arising from dampness or extremes of temperature. Adequate lighting is necessary, since the store must be considered as a working area and poor illumination is hazardous and may lead to labels being misread. In order to make the maximum use of natural lighting, the store should be provided with windows, generally fitted with obscured glass. Grilles may need to be added for security reasons. The space between shelving units must be wide enough for the safe movement of goods and trolleys as well as personnel. The units themselves must be stable, preferably fixed to the walls or floors.

As all goods entering the store must ultimately leave it, there must be easy access to the rest of the building and a lift should be provided if the establishment has more than one level. A small office will probably be required, or at least a desk and filing cabinets for record-keeping purposes.

The entire store should, of course, be made secure against intrusion and have access only to those persons authorized to enter. Since much of the contents are likely to be combustible, suitable fire-fighting equipment must be provided and an automatic fire alarm is desirable to safeguard the premises when unoccupied. Water sprinklers are not recommended as they may cause more damage than the fire itself. An automatic CO_2 system with suitable safeguards, although more expensive, is far more satisfactory.

For moving goods within the store and to the laboratories, suitable trucks or trolleys must be provided. These should be of robust construction, as they invariably receive more rough handling than one anticipates, and must be of suitable dimensions to pass between storage racks, through corridors and doorways and to fit into the lift where necessary. Ladders and steps are required to reach the higher shelves and these should be constructed from tubular steel or aluminium, with non-slip treads, and be maintained in safe condition.

In a multi-storey building there is a risk of flooding from rooms above the store and it is vital to ensure that this risk is reduced to an insignificant level.

Facilities are required for unpacking consignments and checking deliveries prior to transfer to the appropriate section of the store. Deposits are required on some packing cases and if these can be unpacked rapidly and returned to the driver, a considerable amount of effort and expense can be avoided.

The main body of the store is intended to accommodate general stocks such as routine chemicals in the low hazard category, glassware and disposable items, which should be kept in their original cartons where possible. Hazardous goods should be transferred to special areas which will be discussed in some detail later.

STORAGE OF CHEMICALS

Whilst most laboratory chemicals are supplied in small bottles, a number are required in large quantities and are generally much cheaper to buy in bulk packs, ranging from 3 kg bottles to drums and plastic sacks of up to 50 kg. Bottled chemicals are best stored on shelves, preferably of metal construction with a corrosion-resistant finish or of unpainted wood (*Figure 4.3*). They should be lipped at the front edge to prevent bottles falling off if the shelves are overcrowded. Where possible, they should not be situated higher than eye level.

At one time it was fashionable to segregate organic chemicals from inorganic but there appears to be no justification for this system. An alphabetical arrangement is adequate, although this can become somewhat confusing in cases where a compound has more than one name, e.g. ethylene-diamine tetra acetic acid is also known as sequestric acid and versine. It is a good plan to adopt the nomenclature used by one of the main laboratory chemical suppliers and to keep a cross-index on file cards available in the store. Chemicals should be segregated according to their particular properties in order to avoid uncontrollable reactions between incompatible materials in the event of breakages or leakages. Excessive heat or direct sunlight may cause rapid decomposition or evolution of gases from some materials and these possibilities must be taken into account when selecting storage locations for them.

Figure 4.3. Purpose-built storage units. A view of a storage area in a perfumery laboratory showing specially designed units with tiered shelves for rapid location and identification of bottles. The units were designed and built by A.R. Hoare & Co. Ltd (courtesy A.R. Hoare & Co. Ltd)

Large containers, such as drums and sacks, are best stored at low level to reduce lifting but should be kept clear of the floor in case of liquid spillages or flooding. A system of duckboards or low shelving about 10 cm above floor level is suitable.

Many chemicals, particularly those used in the biological sciences, are labile and must be stored at low temperatures, for which refrigeration is necessary. Four distinct storage temperatures are recommended by the manufacturers for different classes of compounds, namely +4°C, −20°C, −40°C and −70°C, and the appropriate refrigeration equipment will be required. Fortunately, these materials are seldom purchased in large amounts and only small units are necessary. Materials in this category should be despatched to the users as soon as possible after arrival, but there are occasions when they must be stored; for example, over the weekend or during the temporary absence of the person who ordered them.

HAZARDOUS MATERIALS

Corrosive chemicals

Corrosive substances, such as concentrated acids, should be housed in a separate area away from other materials. This area should be cool and well ventilated and fitted with substantial shelving which should not be higher than eye level. Small quantities may be stored in ventilated cupboards

Figure 4.4. *Storage of noxious materials. Ventilated storage cupboards fitted at the sides of a modern fume cupboard. A separate extract system is incorporated for the cupboards and runs continuously (courtesy A.R. Hoare & Co. Ltd)*

(*Figure 4.4*). Neutralizers such as sodium carbonate and absorbents such as dry sand should be provided in case of spillage, and protective clothing should be available for use in the event of an accident. Bottle carriers must be used for transporting these materials.

Poisons

Most chemicals in common use are toxic to a greater or lesser degree, but from a storage point of view it is clearly not practical to consider all chemicals as poisons. Highly toxic materials and those covered by legislation[2-4] should be kept in locked cupboards with limited access and only issued to authorized persons against signed requisitions. Problems associated with the issue and use of poisons arise mainly in teaching laboratories, especially in schools where young people can be put at risk, and it is the responsibility of the teacher or lecturer to ensure that such materials are used correctly. In a properly supervised store they should not present any difficulties.

Flammable solvents

Local authority and statutory regulations control the design and construction of stores for flammable solvents which must be inspected and approved before being put into service[5-7]. They are generally required to be separate from the main building and constructed in such a way that a fire will be contained for a specified period of time. A typical design is of brick or concrete construction without windows but with ventilators at high and low level. The ventilators must be covered with wire gauze and should be on at least two sides of the room. The door should be provided with a sill, thus making the floor into a tray, the capacity of which must be at least twice the total volume of solvents stored at any one time. Light fittings and switches must be of an approved spark-proof design, adequate fire-fighting equipment must be provided outside the store and 'No Smoking' signs must be displayed. The door must be of an approved fire-resistant construction and shelves must be of non-combustible materials.

There are many acceptable variations on the basic requirements but the aim should be to provide a store in which it is virtually impossible to start a fire, would contain a fire if one did occur and would not be affected by a fire in another part of the building. Building research organizations continually investigate new designs with these objectives in mind and various ideas have been tried out. These include such details as sand-filled roof voids which collapse in the event of a fire and lightweight roof structures which, in the event of an explosion, enable the expanding gases to escape vertically, thus preventing wall collapse and horizontal blast.

Under UK legislation, certain solvents of petroleum origin are covered by the Petroleum Consolidation Act 1928, together with subsequent amending legislation, and any store carrying these items must be licensed and available for inspection. The amount carried is limited by this licence and it is an offence to store more than this amount. If duty-free alcohol is kept in the solvent store, it is necessary to provide a locked cupboard within the store solely for this purpose. Flammable liquids must not be transferred from one container to another within the store — such operations must be carried out elsewhere.

Gases under pressure

Gases in cylinders are usually supplied at about 150 kgf/cm^2 pressure. This figure can rise very considerably at elevated temperatures and the need for keeping cylinders cool must be obvious. Even at normal ambient temperatures, the pressure is sufficient to cause considerable havoc if the valve is damaged. There have been many cases in which the rapidly escaping gas has propelled cylinders at high speeds for considerable distances and even driven them through brick walls. In view of the possible hazards, it is advisable to provide separate storage for cylinders, further segregation being necessary for flammable and non-flammable gases. Where possible, full cylinders should be separated from empties.

Cylinders should be stored in a vertical position and secured in such a way that they cannot fall or be accidentally knocked over. Straps or chains are suitable for this purpose but it is essential that the system does not release more than one cylinder at a time. This can be achieved, for example, by using a short length of chain for each cylinder rather than one continuous length for securing several cylinders.

In view of the possibility of gas leaks, the store should be well ventilated and light fittings and switches must be of spark-proof design.

It is inadvisable to keep cylinder keys and regulators in the store as there is a temptation to test cylinders for content in the store, before removing them to the laboratories.

Any damaged cylinders should be labelled accordingly and returned to the supplier as soon as possible. Leak testing can be carried out with soap solution or liquid detergent; on no account should a flame be used for this purpose.

Gas cylinders are usually very heavy, so suitable trolleys or cylinder skates must be provided for their transport and some arrangement made for these to be returned to the store after use.

Cylinders should be dated on arrival and used in rotation. Suppliers will nearly always levy rental charges on cylinders and they should therefore be returned as soon as they are empty. Some suppliers are reluctant to supply a full cylinder if no empty one is available for return. If a cylinder is no longer required, even though not empty, it should either be emptied under safe conditions or be clearly labelled as not empty before being returned to the supplier.

Liquefied gases

Liquefied gases not under pressure are mainly used for cryogenic purposes and are stored in insulated containers. It is usual practice to arrange for one's own container to be filled from a delivery tanker in the case of liquid air or liquid nitrogen, an operation which should be carried out in the open but under cover. Containers must not be sealed and must be stored in a well-ventilated area. Liquid nitrogen, although non-toxic, will produce large volumes of gaseous nitrogen which will displace the air in a room, thereby reducing the oxygen content to a dangerously low level. If it is necessary to pour liquefied gases, the reader is referred to Chapter 7 regarding the precautions to be taken. Suitable protective clothing should be available in the storage area.

Liquid ammonia is supplied under pressure in cylinders fitted with a dip tube which permits the removal of liquid under pressure or, by inverting the cylinder, gaseous ammonia. This gas is not only highly toxic but forms explosive mixtures with air and should therefore be treated as a potential hazard and stored only in a well-ventilated area.

Solid carbon dioxide

This substance is also used mainly for cryogenic purposes and should be stored in insulated and vented containers to minimize loss by evaporation. Under no circumstances should it be stored in cold-rooms. It can cause very severe low temperature burns and insulated gloves must be available for persons handling this material. As its storage life is short, it should be ordered for delivery on the day it is required but in organizations where consumption is high, it is usual to arrange for deliveries to be made at regular intervals, say once or twice a week.

Radioactive materials

It is not good practice to keep radioactive substances in the store and they should be despatched to the user immediately on arrival. If, however, there is any reason for retaining such materials in the store, a lockable metal safe should be used, suitably labelled, and the materials kept unopened in their original packages. The local Fire Authority should be informed of the fact that radioactive material may be present in this area[8].

STORAGE OF APPARATUS

There are a number of scientific establishments in which the main bulk of apparatus in stock comprises a range of glassware and disposable items.

Whenever possible such items should be stored in their original packages. Not only does this minimize breakage but it simplifies issuing and stocktaking and is convenient when stocks have to be moved such as when reorganizing or redecorating. Packs should be dated on receipt and used in rotation, although this has the disadvantage that the oldest pack is invariably at the bottom of the stack. Cartons should neither be stored directly on the floor where they may be damaged by spillage or flooding, nor should they be stacked so high that excessive weight is applied to those at the bottom of the pile. If packs are opened and partly used, the labels should be altered as appropriate. Loose items of glassware may be stored on metal shelf units with vertical dividers to separate items into logical groups.

Glassware is easily damaged if items are allowed to touch each other. For example, small barely visible cracks may form which render the glassware unsafe to use, particularly under vacuum. This applies especially to large flasks and precautions must be taken to prevent this occurrence. It is not good practice to accept glassware back into stock once it has been issued, except in the case of items of a complex or costly nature.

Disposable plastic items generally have long shelf lives, particularly if made of polythene or polypropylene. Polystyrene, on the other hand, deteriorates rapidly in the presence of organic solvent vapours and is sufficiently brittle to be cracked by impact, even if kept in its delivery cartons.

Many disposable items are available in sterile packs and these must be issued intact and not opened in the store. Not only could the resultant contamination ruin experimental work but, in the case of syringes, needles, etc., it could cause infections.

Disposable paper products such as tissues and face masks deteriorate rapidly in damp conditions and as these are frequently stocked in large quantities it may be worth considering setting aside a particular area for their storage.

The word 'disposable' implies a low price but a study of current price lists suggests that considerable sums are spent on these items, particularly as large purchases are encouraged by the quantity discounts offered by the suppliers.

Other apparatus

The store may be called upon to carry stocks of general and specialized apparatus, ranging from small electronic components to relatively large items such as water baths and stills. Major items of capital equipment are not normally held in stock, these being supplied directly to the laboratories through the normal buying procedure. Depending on the varied nature of the stock, a range of storage units will be required. It is usual to purchase these as ready-made units but they can be supplied to suit particular requirements. For small items, drawer units are convenient as they are easily subdivided into sections. Shelves and bin units are available and lockable cupboards may be used for items which are liable to be pilfered.

As a general rule, heavy items should be kept near floor level and lighter goods on the higher shelves. This not only reduces physical effort but minimizes the risk of injuries caused by falling objects.

Apparatus with detachable parts and accessories can present problems because there is a tendency for these to become separated, and as some smaller parts are not easily identified they may even be disposed of. It should be possible to keep all such accessories in boxes or bags taped to the apparatus to which they belong.

Photographic and photocopying materials generally have a limited shelf life and should, therefore, be issued in rotation. They are also affected by heat, humidity and chemical contamination and by radiation from ionizing sources, both radioactive and electrical. Photographic film, especially colour material, deteriorates rapidly at high temperatures and in humid atmospheres. The principal manufacturers issue guidance on the storage of their products in tropical climates.

Documentation

The paperwork concerned with buying is discussed in Chapter 3. Stores documentation performs a number of separate functions: it serves as a check on what is in stock and what needs to be ordered, it indicates where in the store a given item may be found, it advises the stores staff of what goods are required to be issued, and is the basis of the accounting system by which the various departments are charged for goods received.

Goods are ordered from the stores by means of a written requisition which must be signed by an authorized person (frequently a head of department or a deputized member of the staff). Collection may be made from the store or some form of delivery service may be provided, depending on the availability of staff and on the amount of work involved. In either case, the person receiving the goods should sign for them and note whether the order is complete or if some items are to follow.

Within the store some form of stock check is required; stock cards perform this function very conveniently as they give all the relevant information concerning each item held. They can be produced to show any information required — a typical example is given in *Figure 4.5*. This card indicates the location of the item in the store, the company from whom it was purchased and the maximum stock to be held, and informs the storekeeper at what level he should reorder.

Nomenclature	Flasks, cmical 500 ml			Location	68
Min. stock	60		Max. stock	240	
Supplier	Blasys & Co.	Cat. No. 2102/5		Std. pack	60
Date	Recd.	Price	Date	Issued	Stock
16—9—77	240	30	19-9-77	20	220
			26-10-77	60	160
			18-1-78	100	60
22-1-78	180	32			240
			11-2-78	24	216
			20-2-78	30	186

Figure 4.5. Typical stock card giving all the relevant information concerning one particular item

The pricing column enables him to make the correct charge when goods are issued and enables him to follow changes in price. Thus, in the example shown, earlier stock is charged out at the old price and later issues charged at the new price. The stock column tells him exactly how many are currently in stock.

It is, of course, essential to the successful operation of the system that the appropriate entries are made on the cards when any issue is made. This is a time-consuming exercise for which there are several alternatives. The simplest, though most costly in terms of capital outlay, is to computerize. This would not be worth while in a small store but if a suitable computer were available within the organization, it could be used on a time-sharing bases (see Chapter 9). A large store could justify the cost of rental or purchase of its own computer, which would be programmed to give all information required and to produce budget figures for the expenditure to be charged to each section. The normal practice is to allocate code numbers for each item, together with a separate code to identify the section to which the goods are issued. Thus, all the operator needs to do is to enter the code number of the particular apparatus, the number of items issued and the code number of the budget to which they are to be charged. The computer calculates the total cost and adds this to the expenditure previously entered. If required, it can be programmed to print out statements at regular intervals, to advise when reordering becomes necessary and to signal that a particular budget is in danger of overspending.

A less expensive alternative is to utilize a system not unlike that used in supermarkets. The staff take goods from the shelves and thence to a 'checkout' where their 'purchases' are recorded. Costing and stock level adjustments are made by the stores staff as a separate exercise. This system has the advantages of keeping paperwork down to the absolute minimum and of accelerating the actual process of issuing goods.

Documentation is further reduced by not actually counting goods in stock but by carrying out visual checks. When stock of a particular item in a bin or on a shelf is below a certain level, an order is placed to bring the stock up to an agreed level. It is worth noting that this system is used with considerable success by a number of large retail organizations and by many large industrial stores.

An adaptation of this system based on the 'cafeteria' principle is particularly useful in educational establishments where materials are issued from the laboratory store to students doing practical work. The apparatus and materials required for each student during a practical period are assembled by the laboratory staff on trays or in boxes and the students draw these from a serving counter. At the end of the period, they are returned, checked and made ready for the next group of students. Using this method, the student has the maximum possible time made available for his work and the laboratory can be made ready for the next class with the minimum of delay. It also obviates the need for the provision of student lockers in the laboratory, since these have been a perennial source of frustration and conflict among both students and staff.

It is of considerable help to the laboratory staff if they are made aware of what apparatus and materials are available from the store. Initially, a stocklist can be prepared and circulated to the various departments concerned. At regular intervals, say every six months, supplementary lists are issued indicating any additions or deletions which have been made.

Some organizations require the value of stock held to be assessed annually for audit or insurance purposes. This is particularly required in commercial firms because the materials in stock form part of the company's assets. In a well-run store, this does not present a major problem for the staff as the records should provide an accurate account of the stock value at any given time. For practical reasons, small numbers of low-value items may be ignored for this purpose and old items which have been returned to the store are generally considered to have no commercial value.

Major items of capital equipment in the laboratories are also included in a stocktaking exercise and assessment of their value can be a fairly complex procedure. The actual price paid is no indication of their present value as factors such as age and condition may reduce this considerably and allowances have to be made for depreciation. The simplest way to arrive at a fairly accurate estimate is to consider the life expectancy of a given item and to devalue it accordingly. For example, if a life of 10 years can be expected in normal use one can devalue by 10% each year; similarly, a five-year expectation will bring about a devaluation of 20% per annum.

Although inflation makes it impossible to replace apparatus at its original price, its true value will not be affected. It is possible to insure for the replacement cost of any item but the premiums are of necessity very high and one would only consider this type of policy for selected items, particularly those which may be used outside the laboratory and which are, therefore, more likely to be lost or damaged beyond repair.

REFERENCES

1. The Nuffield Foundation, Division for Architectural Studies, *The Design of Research Laboratories*. Oxford University Press (1961)
2. The Poisons Act 1972. HMSO, London
3. The Carcinogenic Substances Regulations 1967. HMSO, London
4. The Dangerous Drugs Act 1965. HMSO, London
5. The Petroleum (Consolidation) Act 1928. HMSO, London
6. *The Storage of Highly Flammable Liquids*, Guidance Note CS2, Health and Safety Executive; HMSO, London
7. The Highly Flammable Liquids and Liquefied Petroleum Gases Regulations 1972. HMSO, London
8. The Radioactive Substances Act 1948. HMSO, London

FURTHER READING

Guy, K., *Laboratory Organisation and Administration*, 2nd edn. Butterworths, London (1973)

5
Laboratory Administration

Technical information

One of the major factors that will contribute to the operational success of nearly every laboratory is the ability of its staff to obtain relevant and accurate information quickly and to store it in such a manner that it may be retrieved when required with the minimum of delay. Many of the larger scientifically based organizations employ information scientists who are given the task of gathering and distributing various types of information for the research and other staff. Information retrieval is now a major service and in recent years it has become an extremely valuable asset to those concerned with research and development. However, the mere accumulation of a vast store of factual information or research data is of little use if it is not efficiently stored and made readily available to those who can make use of it. A book misplaced on the shelves of a large library is rendered almost useless to those who may wish to consult it.

Before deciding on the most appropriate storage system, it is useful to consider the type and amount of information that has to be stored. It is also desirable to consider the possible expansion of the system at the same time, so that it will be able to take account of future developments.

In many fields of research, data is generated at such a rate that it can only be handled and stored with the help of a computer. Indeed, many current research projects would be difficult if not impossible to carry out without the assistance of some form of computer facility. At a lower level, much technical information takes the form of laboratory notebooks, catalogues, sales and advertising literature, plans, drawings, scientific journals, textbooks and reference books.

Apart from the occasional use of microfilm recording systems, much of this type of information is stored in the traditional manner, e.g. in bookcases, on shelves and in filing cabinets and cupboards.

Some information, especially if it involves experimental results, may have been gathered at very high cost. In these circumstances, it may be necessary to ensure that it is stored under conditions where it will be safe from deterioration and secure from fire and theft.

It is imperative that all information is adequately catalogued or indexed and cross-referenced so that it may be readily found when required. Cross-referencing is particularly important. Nearly all laboratory workers will have experienced the agony of searching through chemical catalogues and reference works for details of a particular substance, only to find after hours of fruitless searching that it is listed under a totally unexpected heading.

The standard international systems of chemical nomenclature are generally of considerable help when compiling or consulting lists of chemicals, but they are not without their own characteristic traps or disadvantages. The avowed aim of these systems is to ensure that all users and writers will refer to chemicals and their formulae in a commonly accepted, regular and systematic manner. Some of the names classified in this way are long and tortuous and it is sometimes extremely difficult for any person other than a skilled chemist to remember them or to link them with the commonly used names. In these circumstances it is useful if the substance is listed not only under its 'correct name' but also cross-indexed under its various trivial or common names.

If the laboratory manager seeks to provide informed leadership for his staff he will need to devote a proportion of his time not only keeping abreast of the literature but also attending symposia and seminars. Considerable judgement will have to be exercised in deciding which meetings are worth attending and which can be avoided. It is not always the most obvious meetings that yield the most useful information. The laboratory manager should also aim to spend an hour or so each week in the library for the purpose of scanning current journals and abstracts and reading book and equipment reviews, etc. Not only should the text of journals be examined but some attention should also be given to the advertisements and announcements. In this manner much potentially useful information concerning new equipment and techniques may be accumulated.

Many scientific libraries will undertake to supply information to individuals on a regular basis as part of their current awareness programmes. The user is asked to specify his fields of interest and the library then provides relevant headings or abstracts of papers. More often than not this service is backed up with a reprint and photocopying facility. Clearly, a comprehensive service of this nature is expensive to set up and run and can only be provided in large organizations and generally where there is some kind of computer facility.

The laboratory manager also needs to devote some of his time to making and developing personal contacts with other institutions, organizations and suppliers, etc. Allocating valuable time for interviewing representatives from the various firms with which he transacts business may at first seem wasteful, but it may often yield a positive benefit by providing information that would not otherwise be obtained. These personal contacts, often developed over many years, quite frequently provide useful information concerning new equipment and materials under development. Occasionally they result in introductions to other users of the companies' products where the exchange of user information may be beneficial. Personal contacts may also be extremely useful when a product is in short supply or is required very urgently. It is much easier to seek special consideration from a company whose representative has been seen regularly over the past few years than from a company with whom there has been little or no personal contact.

FILING SYSTEMS

Efficient filing systems are important to every organization. They range from the shelf of catalogues arranged roughly in alphabetical order and irrespective of size or shape, to the very elaborate systems used by some large organizations such as government departments.

Good filing systems should be (a) suitable for the type of information which is to be stored, (b) simple, so that all users may readily comprehend and appreciate their working, (c) accessible, so that material may be rapidly retrieved, and (d) adaptable, so that they may be easily altered to take account of changing needs. The system must also allow for the regular review of all material so that obsolete material may be considered for disposal. Often it may be extremely difficult to judge if material is obsolete and can be safely discarded or if it is obsolescent and has to be preserved against a possible future use. It should be remembered that, although space is costly, information once discarded may be impossible to recover and therefore all possible future uses and developments must be considered before it is destroyed.

Some information, such as statutory records and information concerning patients' medical history and treatment, has for legal reasons to be kept indefinitely. Rather than clutter the current files with subject matter that is only rarely referred to, it may be appropriate to store this information elsewhere as archive material. This material is usually accommodated in less expensive and often less readily accessible space but it is important to ensure that the information will be safe and secure. Damp rooms or basements liable to flooding are not suitable. It is also important that all archive material is properly indexed or catalogued so that it may easily be retrieved.

INDEXING SYSTEMS

Two basic systems are commonly used: the alphabetical and the numerical[1].

Alphabetical systems

In these systems, the material is arranged in alphabetical order according to certain commonly accepted rules. The most widely encountered example is the listing of subscribers' names in the telephone directory. The system is most effective when all the items to be filed are similar in size and shape or are contained in standard folders or boxes. This method is generally preferred where the file headings are names, especially where the total number of names involved is not likely to be very large. If, however, there is a sizeable number of names to be accommodated then a secondary and, if necessary, a tertiary unit of classification may be needed. One very large hospital has adapted a system in which patients' records are filed under surnames and forenames, followed by the maiden name of each patient's mother. This very easily surmounts the difficulties encountered in communities where there may be numerous Smiths or Joneses with similar forenames, or where there are large immigrant populations.

Files may also be arranged alphabetically under subject headings or titles. In this case, the key word from the subject or title is taken first, followed by any other subsidiary words that are considered necessary to differentiate the subject. This system is most effective where the title is precise, such as with laboratory equipment or materials, but it may be less so where the title is less explicit.

Numerical systems

These systems are very adaptable and are generally used for ongoing work where new files have to added regularly or where the old files have to be expanded.

The primary reference is the file number, and all material originating from or coming into the file should bear this number. Apart from its number, the file cover usually carries its title and any subtitle, together with the numbers of any files in the collection whose subject matter is related.

A separate card index is used as a key to the system and reference is made to this to find the number of the files to be consulted. The key cards, which are kept in alphabetical order, are headed with the file title and give the reference number, the starting and closing dates and the reference numbers of any related files. Reference numbers are usually allocated serially so that blocks of numbers will correspond roughly with the file starting dates. Occasionally, blocks of numbers are reserved for broad subject groups or particular topics. The use of a separate card index as a key to the system makes it particularly easy to cross-reference the subject matter, as each cross-reference merely entails the preparation of an extra card. When a particular file or folder is full, a new one is commenced which is given the same number followed by a suffix, for example 3001-1, 3001-2, or 3001 Vol. I, Vol. II, etc.

DEWEY DECIMAL CLASSIFICATION

This system is widely used in libraries for the classification of books. The 10 main groups are numbered 000 and 100—900 inclusive and these are each divided into nine subgroups, 210, 220, 230, etc., each of which may be still further subdivided into smaller groups, e.g. 201, 202, 203. If further division and subdivision is necessary, two places of decimal points may be used. The specific numbers allocated to the various subjects as used in libraries follow a generally accepted standard but the system may be readily altered to suit any given set of circumstances. The system lends itself well to the classification of scientific and technical information but considerable care must be taken when allocating the major divisions. Once the primary groups have been decided upon, the system becomes capable of considerable expansion.

Laboratory records and record-keeping

Laboratory records may be divided into two groups: (1) those which have to be kept to comply with the law, and (2) those which, although there may be no statutory requirements to keep, will nevertheless help to ensure the smooth and efficient running of the laboratory.

The maintenance of accurate records is an essential part of modern laboratory management and although much of the day-to-day work may be delegated to subordinate staff, the manager cannot avoid the responsibility of making certain that the work is properly organized and carried out. He must also ensure that all the legal requirements, including the making of the appropriate returns, are correctly observed. With the increase in legislation affecting laboratory work and with the constant necessity of keeping costs under review, it is essential that all staff are aware of their personal obligations under the various statutory provisions. Furthermore, they must be made aware of the possible consequences arising from failure to keep the appropriate records.

Many of the records that have to be kept have been dealt with in the appropriate chapters but it may be useful at this stage to list some of them in order to see what is expected of the laboratory manager:

(1) Accident books and records of dangerous occurrences.
(2) Ethyl alcohol, its purchase, stock and use.
(3) Dangerous drugs, purchase, stock and use.
(4) Radioactive substances, purchase, stock, use and disposal.
(5) Experimental animals, purchase, stock, sickness, use, etc.
(6) Experimental animals, personal records of experiments, joint experiments.
(7) Financial records: (a) budgets, financial allocations, grants; (b) orders, deliveries, invoices and accounts.
(8) Building maintenance: (a) internal and external; (b) planned preventive maintenance.
(9) Equipment maintenance, servicing and replacement.
(10) Insurance of major equipment, insurance inspections.
(11) Equipment inventories.
(12) Equipment loans and transfers.
(13) Library books, stock, loans, acquisitions.
(14) Personnel records: (a) sickness and absenteeism; (b) vaccinations, immunizations; (c) X-ray examinations; (d) exposure to specific hazards — radioactive substances, carcinogens; (e) salaries, promotions, awards, etc.; (f) misdemeanours, disciplinary action.

Office facilities and equipment

Almost every member of the laboratory staff will need to be provided with space and office facilities of one kind or another to enable him to record and present the results of his work. The provision of major equipment such as photocopying machines and filing systems has been mentioned elsewhere in this book, but consideration must also be given at the planning stage to the provision and location of the basic equipment and services that will enable the laboratory staff to perform their office and administrative tasks efficiently. Fully serviced laboratory space is expensive and clearly it is wasteful to use such space to accommodate desks and other office furniture.

It has become the custom in recent years to design and build laboratories that are approximately twice as deep as they are wide, with the main doors at one end and the windows at the other. Benches are usually arranged almost continuously along the longer walls. It has become the almost invariable custom to position a free-standing desk under the window. In theory, this may seem perfectly reasonable but its practice introduces a number of difficulties. If proper access, for example, to the cupboards under the bench adjacent to the desk is to be retained, then the distance between the benches needs to be increased and further complications will ensue if the fire-prevention authorities insist on the provision of an escape route at the far end of the laboratory (*Figure 5.1*).

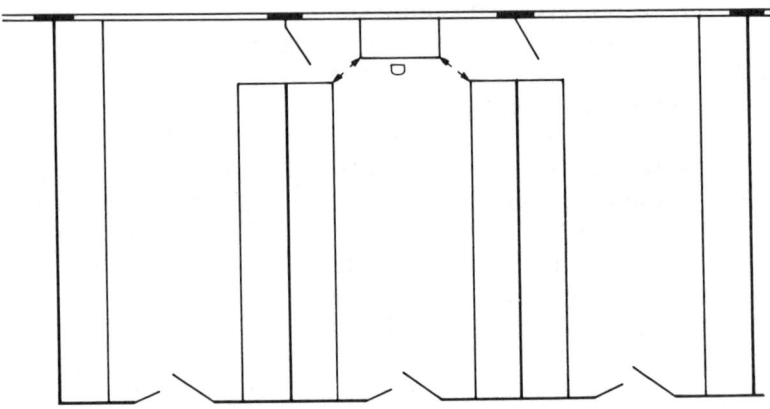

Figure 5.1. Obstruction of fire escape route showing insufficient width of route when a desk (D) is placed under the window

Junior members of staff may require only limited writing space in the laboratory in which to record the results of their experimental work. This may be provided by making one section of the bench with a working surface approximately 76 cm above floor level.

Intermediate grade staff will usually require writing desks and filing facilities in the laboratory so that they may keep a watchful eye on the laboratory work whilst attending to their clerical duties. With the increasing use of pocket calculators and portable recording and dictating equipment, both junior and intermediate staff will require cupboards and drawers that can be securely locked.

The more senior staff will require office accommodation that affords them a degree of privacy to enable them to perform their administrative duties free from interruption and in relative peace and quiet. It is an unfortunate fact of life that, as one rises in the laboratory hierarchy, one is forced to spend more and more time away from the bench for the purpose of either compiling reports or reading the reports of others.

Many modern laboratory buildings have been designed with small offices or 'think boxes' adjacent to the laboratories, so that section leaders, etc., are not too far removed from their work or from the staff for whom they are responsible (*Figure 5.2*).

Typewriting facilities are usually provided for the more senior staff, although many staff now type their own reports.

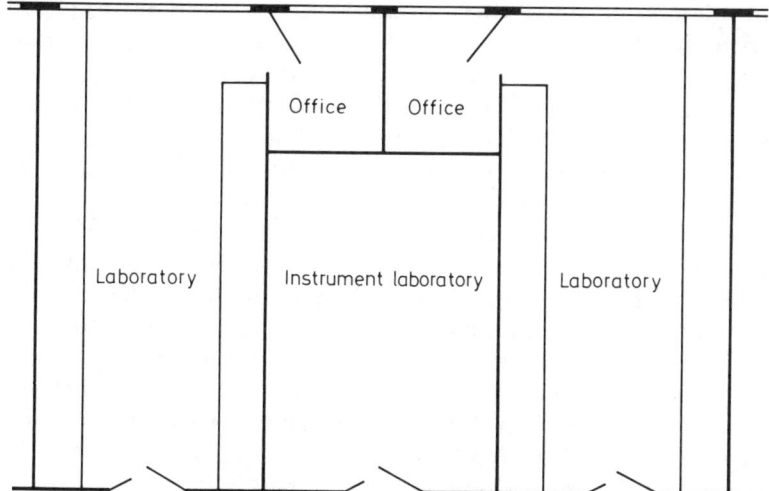

Figure 5.2. Provision of laboratory offices showing how two small offices may be formed to serve two laboratories. The major part of the centre module is used as an instrument laboratory or as space to contain a cold-room, autoclave room, etc.

Where the typing staff work in a pool, it may be found advantageous to allocate certain members of the pool to a specific group of staff. In this way, the personal idiosyncrasies of staff, such as almost indecipherable handwriting, may be more efficiently dealt with.

Decision-making

In laboratory work, as in everyday life, staff are constantly faced with the necessity of making decisions. Sometimes these are simple and require little thought and are arrived at almost unconsciously. Sometimes, when much is at stake, their resolution may require a great deal of thought together with the exercise of considerable discretion and judgement. It is doubtful if judgement can be taught, but training and disciplined thinking can do much to ensure that reasonable decisions are made without undue delay.

Before any decision is made, the problem must be identified and stated in clear terms. Often in the rush and anxiety of getting things done, the real problem is overlooked or incorrectly identified, and one can hardly complain if the results are disappointing. Having determined the problem, one must then assemble the relevant information with the full appreciation that this may not be readily available and that further enquiries may have to be made. The next move is to consider the various options open to the solution of the problem. Some options will be too costly, impossible to achieve, or illegal and may therefore be discarded. The remainder should be arranged in order of priority after considering all the possible results of each option. It is probable that each option will have both plus and minus points and these will have to be traded against each other in determining the most appropriate solution.

Having reached a decision, it will be necessary to put that decision into effect by informing those concerned without delay. Once the decision has been given or carried out, the manager must be prepared to accept responsibility for its consequences. Occasionally new factors will come to light after the decision has been made which will necessitate a change of mind. In these circumstances, all the evidence should be reconsidered before announcing the change of plan. It is important, if the manager is not to be thought of as 'blowing with the wind' or, worse still, not to be accused of vacillation, that decisions once given are not changed too often.

The decision-making process may be summed up as follows:

(1) Identify the problem.
(2) Assemble the relevant information.
(3) Consider the options and their probable results.
(4) Arrange the options in order, with their attendant advantages and disadvantages.
(5) Decide.
(6) Accept responsibility for the decision.
(7) Change the decision only if further new facts come to light.

SEEKING ADVICE

Sometimes, through lack of specific experience in a particular field, the laboratory manager will need to seek further information and the advice of others before deciding on the most appropriate course of action. At best, this process is time-consuming and, at worst, it may be an unpredictable gamble. It is rewarding to have one's own opinion confirmed, but a single confirmation does not mean that the original decision is correct. There is always the risk that undue weight will be attached to opinions that closely agree with one's own and that insufficient attention will be given to views that are likely to prove difficult or unpopular to implement. There is also the possibility that the person whose advice is sought will be biased in one way or another. It is unreasonable to expect that a person who has recently spent a large proportion of his laboratory budget on a specialized piece of equipment will say that it is useless or has several major faults. Conversely, if he is campaigning for a different piece of equipment, he may well say that his present model was unsatisfactory from the day that it was bought.

Some people expect that their advice will be followed to the letter and are likely to show considerable annoyance when this does not occur. All advice should be carefully weighed and due consideration given to the experience, status, judgement and objectivity of the giver before making a final decision.

There is an implied obligation to heed advice when it is sought from persons of superior rank and to accept the opinions that are given, especially if these opinions are expressed as administrative principles. More often than not the manager will also be seeking the views of his colleagues and associates in an informal manner to help him reach an overall decision that is based on experience broader than his own.

If there is likely to be a very wide divergence of opinion, or if the subject is particularly controversial, it is important that as many views as are reasonably practicable are taken into account. Consultation is just as important in the laboratory as it is generally assumed to be in politics or public affairs, but at the end of the day it is the manager who has to make the decision and it is he alone who will be held responsible for its inadequacies.

STAFF MEETINGS

If a number of persons are to be asked to express an opinion, it is important that the problem is explained in similar terms to each individual, otherwise the manager will have considerable difficulty, when he attempts to draw together all the strands of his discussions, in arriving at a solution. A more effective but time-consuming method is to call a meeting of all the interested parties so that the problem, together with its alternative solutions, may be thoroughly discussed. Following a meeting of this nature, the manager still has the difficult task of assessing the possibly conflicting advice which has been given. As mentioned above, the responsibility for the decision and for its implementation still rests with him. The consultative process, including the setting up of advisory meetings and committees, is a useful management tool, but it should not be allowed to become an opportunity for sharing or avoiding responsibility for difficult or unpopular decisions.

Informal meetings are usually arranged *ad hoc* to discuss a particular subject. The discussion will be led by the person who has called the meeting and all should be given adequate opportunity to speak and to come back after others have spoken. Only very rarely are minutes or notes of the meeting circulated to the persons taking part in the discussion.

ADVISORY COMMITTEE MEETINGS

These gatherings are usually fairly informal meetings of staff or of experts called together by the director or head of the establishment to consider a particular problem or series of related problems. They may meet once only or on a number of occasions, or they may be a standing committee charged with holding regular meetings to monitor a particular aspect of the organization's work. Safety committees and finance committees are popular examples of meetings falling into this category.

When such a committee is formed or convened, they may be given general or specific terms of reference relating to their objectives and method of reporting, etc. They will be expected to convey their conclusions and recommendations, together with any necessary supporting evidence, to the director by means of a written report. These conclusions and recommendations are usually arrived at by general agreement and only rarely do members find it necessary to vote on a particular issue or to submit minority reports. If the chairman should ask for a 'show of hands', it is usually because he is trying to gauge the strength of opinion for or against a particular course of action. If there is no general agreement, he may decide to continue the discussion until a more uniform opinion emerges. This is known as reaching a consensus.

Having received the report, there is no obligation on the part of the director to accept it or to implement all its recommendations. He may choose merely to regard the committee meetings as a forum for discussion and then proceed to reach a decision on the subject, whilst no doubt keeping the contents of the report in mind.

FORMAL COMMITTEE MEETINGS

These meetings are conducted in accordance with precise terms of reference which define the objectives of the committee and give details of how, when and where the meetings are to be conducted. Members may be either nominated or elected. Occasionally, the chairman may be asked to find, and if necessary coerce, suitable members to join the committee.

In the UK, the rules adopted for the conduct of formal committees are generally based on those developed over many centuries for the conduct of parliamentary business. In local government, Trade Union and other public affairs these rules are often the subject of endless debate and this eventually leads to the drawing up of very precise protocols and terms of reference. In scientific matters it is to be hoped, however, that the subject is considered to be more important than the procedure. For an exposition of the minutiae of committee procedure, the reader is referred to any of the many excellent books that are available, the most widely used of these being undoubtedly *The ABC of Chairmanship* by Lord Citrine[2]. First published in 1939, this book has passed through a number of editions and reprintings and is generally regarded as the Bible of chairmanship.

A formal committee needs to have a chairman and secretary and, in addition, if the work is particularly heavy, a minutes secretary whose task is solely concerned with taking the minutes of the meeting. Sometimes the officers are nominated by the person or body that brought the committee into being; if this is not so, then they are elected by the committee members at their first meeting. Where the committee is permanent and is representative of various disparate groups (e.g. a joint consultative committee of management and staff), it may be agreed that for the first year the staff shall have the right to nominate the chairman and that management shall provide the secretary and that these rights to nominate the officers shall be revised each year.

Before the meeting, each member will expect to receive an agenda which has been drawn up by the secretary in consultation with the chairman. The agenda gives the items that are to be discussed and lists the order in which they are to be taken. After the meeting, the secretary draws up the draft minutes and circulates them to all members, together with a reminder of the next meeting. The first business of the following meeting is to agree the minutes of the last.

The constitution of the committee may allow the chairman or officers of the committee to take any urgent action that may be necessary between meetings, but generally no action arising from committee discussions may be taken until the minutes have been agreed by the full committee and signed by the chairman.

It is important to remember that all the members are bound by any decision which is reached in committee and that an individual who has voted against a particular item may not afterwards publicly dissociate himself from the majority decision.

Where the business of a committee is to compile a complicated code of practice or any other detailed specification, it is usual for the officers or a nominated member or group of members to prepare a draft which is circulated to members prior to the meeting. This will serve as a nucleus for discussion and may considerably expedite the work of the full committee and save much time at the meeting. Sometimes, if the business is very complicated, the committee will refer the matter to a subcommittee for preliminary investigation.

When the main committee has completed its work, its recommendations, etc., are presented to the person or body that was responsible for its formation in report form. In government, major reports are referred to by the name of the committee chairman, e.g. the 'Howie' report, named after Sir James Howie, which is referred to in the References for Chapter 7.

When the report is presented to the person or body that was responsible for the committee's commission, it is normal practice for the report together with its recommendations to be accepted, but this may not always be the case. Clearly, if a number of experts meet to discuss a problem, they are more likely to produce an acceptable and factually correct answer than could be expected from a single non-expert. Although they may be technically correct, they may not be in possession of all the information that is needed to implement their decisions. This often happens with government-commissioned reports, in which the experts produce the answers but the time may neither be opportune, nor may the resources be available, to accept their recommendations.

REFERENCES

1. Guy, K., *Laboratory Organisation and Administration*, 2nd edn. Butterworths, London (1973)
2. Citrine, The Right Hon. Lord, *The ABC of Chairmanship*. NCLC Publishing Society Ltd, London (1968)

FURTHER READING

Banks, A.L. and Hislop, J.A., *The Art of Administration*. University Tutorial Press, London (1961)
Betts, P.W., *Supervisory Studies*, 2nd edn. Macdonald & Evans, London (1973)
Brown, R.G.S., *The Administrative Process in Britain*. Methuen, London (1970)

6
Service Departments and Special-purpose Rooms

In almost every laboratory complex there is a need for a number of general services which provide essential back-up facilities for the work. It is not practicable to consider these as part of any one department or section and although it is usual to make certain members of staff responsible for them, they are likely to come under the general supervision of the laboratory manager.

It is not the intention of this chapter to discuss the various laboratory techniques used in providing these services but to consider the management aspects of their efficient organization and operation.

Glassware washing and sterilizing facilities

Despite the increased use of disposable items intended to replace it, glassware is still used in appreciable quantities in many laboratories, particularly those engaged in chemical or biochemical work. In biological laboratories, there is also a need for glassware sterilizing facilities.

The actual amount of glassware used will vary considerably from one department to another and may indeed vary daily within the same organization. The design and running of the washing system must be such that it can cope with the maximum demand made on it. In the small laboratory, no special facilities may be needed other than one or more large sinks and a drying-oven, but the large complex will require a highly organized unit to handle the substantial amounts of glassware in daily use.

Although one person can process several hundred items in a working day with the minimum of equipment, it is usually necessary to provide automatic washing machinery where large numbers of items are to be washed, particularly if they are heavily contaminated. The capital outlay required to provide such machines is undoubtedly high and they are fairly costly to operate as they use a considerable amount of electrical power and distilled water, but the better machines clean glassware to a very high standard with the minimum of breakages and require very little attention other than loading and unloading. Most of them have a drying cycle and will process a full load from start to finish in about 40 min. Ultrasonic cleaning equipment is also available and is particularly effective for such items as glass syringes which are easily damaged by brushing or high pressure jet washing. Although the capital cost of such machines is fairly high, they are not expensive to run.

Whether glass washing is done by hand or by machinery, the process should be based on a flowline concept; the dirty glassware is brought in at one end of the system and clean items are removed at the other end. Not only does this arrangement make for increased efficiency but it reduces the risk of dirty apparatus becoming inadvertently mixed with clean. *Figure 6.1* shows a suggested layout.

The bulk of the load will be processed in the conventional way and either stored for collection or returned to the section from which it came. There are, however, some items which will require special treatment or which will have received some preliminary attention before being sent for washing. Glassware used for radioactive work should be decontaminated in the laboratory by soaking in a detergent formulated for that purpose, followed by thorough rinsing. Items containing known or suspected pathogens should be sterilized by a method known to be effective, prior to being sent for washing.

Some items may be so heavily contaminated that the cost of cleaning them exceeds their replacement value and these should therefore be discarded in a safe manner, bins being provided for this purpose.

Drying-ovens are more efficient if fitted with fans exhausting outside the building and may be run at a lower temperature than non-exhausted units.

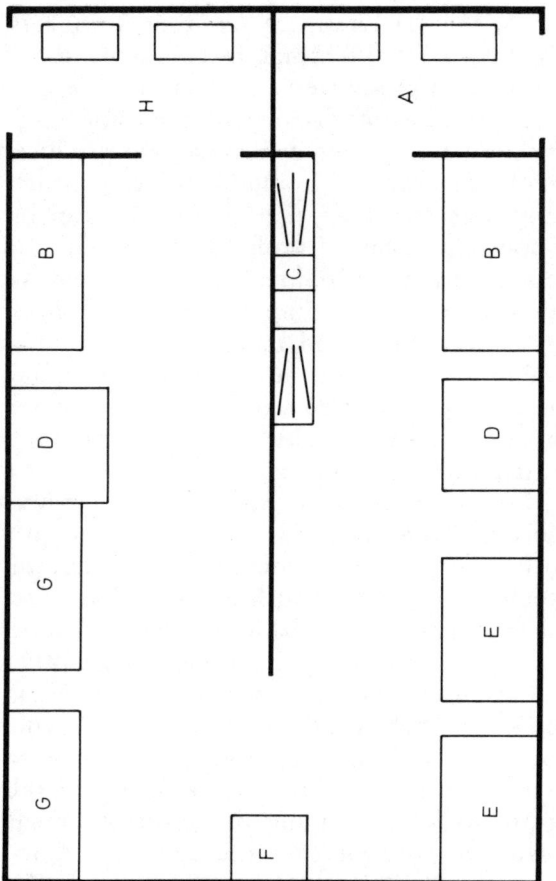

Figure 6.1. Suggested arrangement for a glassware washing and sterilizing unit showing dirty and clean sections: A, reception area for trolleys of dirty glassware; B, sorting benches; C, sink unit; D, sterilizers; E, washing machines; F, still; G, drying ovens; H, collection area for trolleys of clean glassware

Adequate supplies of hot and cold water are required, together with large-bore waste pipes. Distilled water is required in large quantities depending on the size and number of machines installed and is best provided by a still in the washing area from which it can be piped direct to the machines.

A high-current electrical supply must be available to provide power for the machines, drying-ovens, sterilizers and still. Good lighting will aid visual inspection and reduce hazards from cracks in glassware which may not be obvious in poor light. A high level of room ventilation is necessary to reduce condensation and to remove excess heat produced by the equipment. Floor covering should be of a type which can withstand being wet for long periods and be capable of resisting the constant wear of trolleys. It should also be sufficiently resilient to minimize breakage of the items which are inevitably dropped from time to time, and must be comfortable to staff who spend most of the day on their feet. Floor drains are a necessity as the surface is certain to be wet in some areas and will also require daily washing.

Sterilization of glassware may be necessary for two reasons. First, if pathogenic organisms are in use these must be destroyed for safety reasons. Secondly, if glassware is used for biological work, it must be sterilized to prevent carry-over into any other experimental situation, an occurrence which could produce misleading results.

There are several methods available for the sterilization of glassware, some of which are more suitable than others in particular circumstances, and the advice of a specialist in the field should be sought before any method is considered to be satisfactory for a particular organism or group of organisms[1].

Chemical sterilization is suitable in some cases. The process consists simply of soaking the items in a disinfectant such as sodium hypochlorite solution for the requisite time and rinsing with water. Many detergents are themselves effective against some organisms.

Steam will kill some pathogens but, at atmospheric pressure, it is ineffective against a large number. Subjected to 1 kgf/cm^2 pressure in an autoclave, most organisms are destroyed in 15 min, although there are some which are resistant even to this treatment. However, autoclaving is the most widely used method of sterilizing glassware and is totally effective against the majority of commonly encountered pathogens. Where doubt exists, tests should be carried out using a culture of the organism, which is tested for viability after the process has been carried out.

Autoclaves may be heated by electricity, gas or steam, the latter being more rapid in operation but requiring a live steam supply from a suitable boiler.

Dry heat in closed ovens at 160°C is effective in many cases but has the disadvantage that some contaminants may be baked on to the glass and become difficult, if not impossible, to remove. Some items which can be neither heated nor chemically treated can be sterilized by gamma-ray irradiation. It is not feasible to carry out this process in the laboratory but there are specialist organizations which offer this service. As an alternative, ultraviolet irradiation may be effective in some circumstances, but as glass is opaque to the useful spectral band this method is only capable of sterilizing those surfaces which can be exposed to the direct radiation.

Once sterilized, glassware should be labelled accordingly, in order to identify it for future use. In some cases, glassware will need to be sterilized before washing, to avoid contaminating personnel and other glassware, and to be sterilized again after washing prior to being issued for re-use.

The cost of any sterilizing process is inevitably high and should be compared with the cost of pre-sterilized disposable ware, which can be substituted in many instances.

It is clearly impossible to assess the actual cost of washing individual items with any degree of accuracy, but it may be the organization's policy to charge each department for this service. An approximate estimate can be obtained by dividing the total annual running costs of the unit by the number of trolley loads of glass processed, and calculating the charge on a *pro rata* basis.

Radioisotope laboratories

The purchase, use and disposal of radioactive material in the UK is controlled by the Radioactive Substances Act 1960, administered by the Department of the Environment[2].

Premises in which radioactive materials are to be kept or

used must be registered with the Department, which issues a Certificate of Registration that imposes certain conditions on the user. There is usually a limit to the amount of activity which may be held on the premises at any one time and the use of certain isotopes may be prohibited. Separate authorization must be obtained for the storage and disposal of radioactive waste, and the methods of disposal and quantities which may be disposed of in a specified time are also defined.

The Ionising Radiations (Unsealed Radioactive Substances) Regulations 1968[3] specify standards of construction for radioactive areas, together with the provision of specified facilities. There is no legislation controlling the techniques used in the laboratories but two major Codes of Practice exist, one relating to persons engaged in research and teaching[4] and the other to persons employed in hospitals.

The design of laboratories for radioisotope work (*Figure 6.2*) is discussed by several authors to whom the reader is

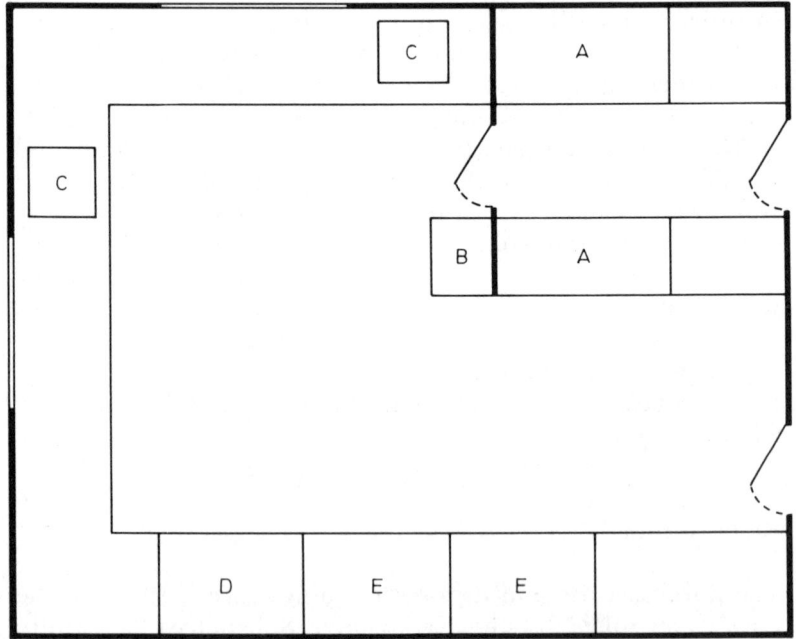

Figure 6.2. Radioisotope laboratory showing area partitioned off for counting: A, counters; B, data processing; C, sinks; D, glove box; E, fume cupboards

referred (see list of Further Reading). The prime consideration is to provide an area which can be easily decontaminated in the event of a spillage and to this end floors and work surfaces should be made from impervious materials free from open joints. Sink waste pipes should be sealed throughout their run to the main sewer, the pipe run being as short and straight as possible. A high level of ventilation is necessary, up to 60 changes per hour in Grade A (High Activity) Laboratories[5], and fume cupboard extracts should be sufficient to provide an air flow of at least 0.5 m/s at the work surface (*Figure 6.3*). The exhaust should be as high as possible and placed so that no emission can fall back into occupied premises, regardless of weather conditions.

Figure 6.3. Fume cupboard suitable for use with radioactive materials. Controls are operated externally; the working surface is easily removed to expose the deep tray base; a pull-out ventilated storage cupboard is fitted beneath the working area and linked to the main extract system (courtesy A.R. Hoare & Co. Ltd)

It may be necessary to use lead containers and screens on the benches and the construction of these must be adequate to carry their weight. Where possible, counting should be done in a separate room communicating with the laboratory.

From the management point of view, the main problems associated with radioactive work relate to record-keeping and safety. The Radioactive Substances Act requires that an accurate account is kept of all radioactive material which is brought into the premises, the amounts disposed of and the method used for their disposal. It must be possible at any time to assess from the records the precise amount of activity currently on the premises. There are, of course, numerous methods which may be used for this purpose, depending on the particular circumstances; one such method is as follows.

At the time of ordering each substance, the appropriate entry is made in a log book and, on receipt, the date and name of the person for whom it was ordered are recorded. It is issued to him with a disposal record form which is completed by the user and returned when the material has been used. The problem is complicated by the fact that the material will probably not all be used at once and is further complicated where several people are using different materials concurrently. It is clearly not possible to maintain records which are accurate to the nearest microcurie, since recovery experiments seldom result in 100% recovery and, with short-lived isotopes, radioactive decay is not easily calculated. Fortunately, the Act does not require such highly accurate recording but it does ensure that radioactive materials are not used and disposed of without consideration for the safety of others and that they are not allowed to accumulate to dangerously high levels. The latter problem sometimes calls for firm discipline, as some workers are reluctant to dispose of residues (it requires extra effort) and will argue that it is being kept as 'it may be needed at a later date'.

The monitoring of personnel is ultimately the responsibility of the laboratory manager, who must also keep records of any exposure to dangerous levels of radioactivity. The normal monitoring method is to use film badges which are issued to each person at regular intervals, usually fortnightly. When new badges are issued, the used ones are processed and an assessment made of the dose received by each user. In the event of a dose measurement being above the acceptable level, the person concerned must be medically examined and not permitted to do any further work with radioactive material until cleared. All doses must be recorded and any employee leaving his job should be issued with a document stating the dose he has received during his employment.

In the event of a spillage, the area should be closed off and decontaminated immediately, using the correct procedure. The persons carrying out this task must be protected with suitable clothing, including gloves, boots, face masks and goggles. The object should be to contain the material rather than to dilute it. If it is a dry substance it should be wetted to prevent it from spreading. This is best done by laying sheets of moistened tissue over the area. Liquid spillages should be absorbed with dry tissues, precautions being taken to avoid spreading. When all surfaces are dried by this method they are then washed with suitable detergents which are in turn soaked up with tissues. All tissues used are placed in double plastic bags and sealed for disposal and the area checked with a monitor[6]. If the area is found still to be active, washing is repeated until monitor readings are down to background level. A record must be kept giving details of spillage, decontamination and monitoring procedures. If it is not possible to decontaminate the area completely, it should be closed off until further action can be taken, calling in specialists if necessary.

Access to radioisotope laboratories should be restricted to those persons directly concerned with the work and if service or maintenance is to be carried out, it is essential to monitor the area and equipment prior to such work being started. Only competent trained personnel may be allowed to work with radioisotopes.

It is sometimes necessary to use radioactive materials in work involving animals, in which case they should be housed in a separate room in the animal complex. Whenever possible, smaller animals such as rats or mice should be used as these can be housed in impervious plastic boxes which contain the waste products and are easily decontaminated.

Animal handlers must be instructed in dealing with waste from the animals; the room, together with cages and equipment, must be monitored to ensure that activity is safely contained. Litter from the room should be disposed of separately and not mixed with general animal house waste. Tissues, carcasses and excreta from the animals are likely to be radioactive and, if required for the experimental work, must only be processed in the laboratory and finally disposed of as radioactive waste.

Photographic units

Photography has become such a comprehensive tool in modern science that photographic units have become an essential part of all but the smallest organizations. The two principal uses for scientific photography are first as part of the investigative work, such as observing transient phenomena which cannot be seen by eye, and secondly, for recording events or data for further study, publication or demonstration.

Photographic materials are available in an enormous variety of types and sizes and it is sound economic policy to standardize, using the minimum range which will satisfy all demands. Most work can be done on 35 mm film stock, with the occasional use of 5 in by 4 in stock where fine detail or giant enlargements are needed. Both these sizes are available as black-and-white and colour materials. The sizes chosen will, of course, decide the format of the equipment required, such as cameras, processing equipment, enlargers, etc., and this decision must be made at the early stages of planning the unit.

Instant picture systems, in which the film is processed in the camera, have many applications in scientific work. Black-and-white or colour prints, black-and-white negatives or positive transparencies for projection can all be readily produced from currently available materials. Future developments will almost certainly include colour transparency materials.

Processing techniques are now so well established that acceptable results can be obtained by any worker of reasonable skill, although it must be emphasized that an experienced photographic technician will produce much superior results in less time and with virtually no wastage. If the demand for photographic work is sufficient to warrant setting up a well-equipped dark room, it is almost certain that one or more photographic technicians will be necessary.

A suggested layout for such a unit is shown in *Figure 6.4* and further suggestions regarding specific designs may be obtained from the major photographic companies.

The amount of photographic work which must be carried out in the dark or under safelight conditions is surprisingly low; for example, in processing films the tank must be loaded in the dark but once this is accomplished all subsequent processing may be carried out in daylight.

Long periods of work in dark rooms can become very demoralizing and it is sound practice to rotate the work schedule of photographic technicians to enable them to spend at least part of their day on tasks which can be carried out in normal surroundings. Such work includes taking of photographs, trimming and mounting prints and slides and preparation of working solutions.

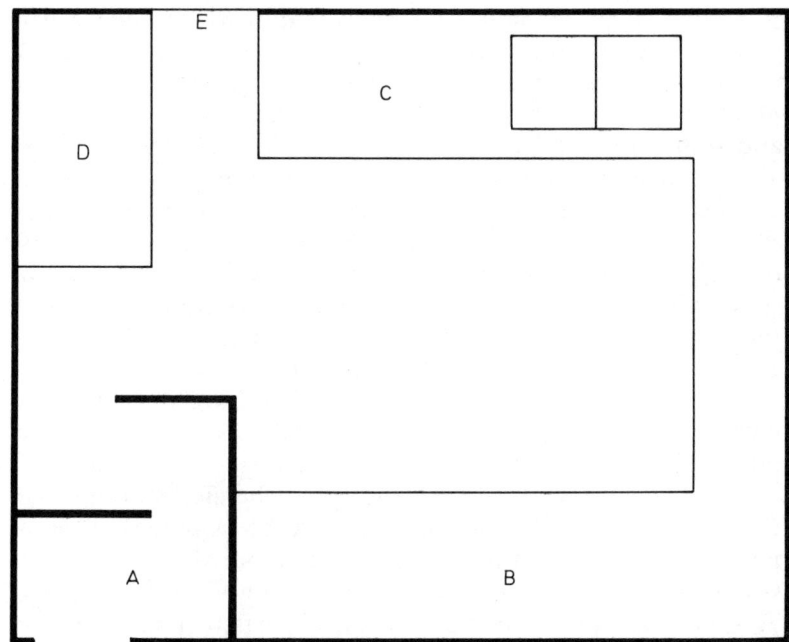

Figure 6.4. Photographic dark room: A, light trap entrance; B, dry bench; C, wet bench; D, storage cupboards; E, emergency exit

The cost of photographic work is fairly high and it is usual to charge the various departments for work which is done. For this reason, it is necessary to keep records of all work, together with the type and amount of material used. As costs continue to rise, a record should be kept of current prices to enable accurate charges to be made.

Exhausted fixer solutions contain a high concentration of silver salts and these should be stored for collection by a silver recovery company, or processed on the premises with a view to recovering the silver. The sale of such residues can provide a useful sum of money to finance some small project or to provide petty cash.

Any work going out of the organization must be carefully checked for quality as it represents a product of the organization from which its standard may be judged by others. A lecturer, for example, whose slides are of a low standard is a poor advertisement for the department which produced his slides.

Safety is an all-important aspect of work in the photographic unit. A considerable amount of electrical equipment is used, such as lighting units, enlargers, glazers and dryers. Much of this is used in conditions which are far from ideal and some are used by persons working alone in closed rooms; all electrical equipment should therefore be inspected and serviced at regular intervals. Some form of emergency signal should be provided so that a person needing assistance may readily attract the attention of others. There should also be some visual indication that a room is occupied, thereby removing the possibility of the occupant being overlooked in the event of an emergency such as a fire.

As with laboratories, all rooms should have provision for escape in emergencies, a fact frequently overlooked in the planning of dark rooms.

When using or handling photographic chemicals, contact of the chemicals with the skin should be avoided. This is particularly important with regard to colour developers, and seamless disposable gloves in good condition should be used. Certain photochemicals present particular hazards and most manufacturers label these products with the appropriate warning.

A high level of ventilation is required in photographic units, especially if processing is carried out at temperatures in excess of $30°C$. Air changes of $8-10/h$ are recommended.

Photographic departments, more than most, require careful work planning to obtain a high level of efficiency. To process 20 colour films, for instance, and then to be presented with another one after they have been finished is clearly an inefficient way of working. All requests for work to be done should be in writing, preferably using a standard printed form, giving clear instructions detailing exactly what is required. It is not uncommon to see such requests as 'Please supply four prints from enclosed negative'. The technician must then waste time establishing the size of print required, whether glazed or matt, single- or double-weight paper, whole negative enlargement or part enlargements, date required and so on.

In the UK, the laws of copyright[7] state that all photographs are the property of the person taking them or of his employer. It is not permissible to reproduce such photographs for sale or publication without authority from the copyright holder and if such authority is granted, this should be acknowledged.

Photographic negatives and prints are often required for future use and should be stored using a filing system which enables them to be retrieved rapidly. The method used will depend on the amount of material involved and on the type of work but the negatives and prints could simply be filed under various headings such as subject matter, departments or names of persons for whom the work was done. Some form of cross-reference is an additional aid to location.

Slides required for lectures are often requested at very short notice and it may be necessary to postpone other work to ensure that these are ready in time. The person whose work is delayed may justly feel annoyed, a situation which would not have arisen had the work been properly planned and requests made in good time. There will, of course, be occasions where notice cannot be given; the very nature of photography makes it an ideal medium for recording unexpected and transient phenomena and some allowance must be made for the worker who needs to record such events.

Cold-rooms

Biological materials may decompose rapidly at normal ambient temperatures and most laboratories handling such materials have a need for one or more cold-rooms in which to store and to work with these substances.

At one time it was usual to provide one large room kept at $+4°C$ and a smaller room at $-20°C$, although in many cases the latter has been replaced by a number of deep-freeze cabinets sited in the laboratories; nevertheless, some organizations still require a small room. Small refrigerated enclosures of suitable sizes are also available for housing such individual items of equipment as fraction collectors in the laboratories.

The presence of cold-rooms poses a number of problems for the laboratory manager who must accept overall responsibility for their use. Like many other multi-user facilities, they are usually neglected by the people who use them on the assumption that somebody else will look after them. Although cold-rooms are intended to be working areas, there is a tendency to use them as low temperature storage areas, with the result that they become cluttered with materials, many of which are useless although the owners do not wish to part with them. A considerable amount of persuasion (and sometimes ruthless discipline) is needed to ensure that the cold-room does not become a dumping ground, and at least once a year the room should be thoroughly purged of obsolete material. Any member of staff leaving the organization should remove his accumulated materials and equipment. It should be made a firm rule that anything deposited in the room must be labelled to indicate who placed it there and on what date. Any item not so labelled should be removed for disposal.

In many organizations, the materials may represent years of work and may have been collected at great expense, perhaps involving trips overseas or collaboration with other laboratories. Any failure of the refrigeration plant could, therefore, result in the loss of materials which are impossible to replace. Although it is not possible to guarantee against breakdowns, they can be minimized by regular maintenance work. This should be carried out, at intervals of not less than six months, by competent engineers, preferably employed by the company who supplied the new plant. Repairs and replacements should be made as soon as possible and, if the equipment is out of action, the room should be sealed, as in this condition it should hold its temperature with only a slight rise for some hours.

Externally visible thermometers and temperature-rise alarms will give early indication of malfunction and, in some cases, a temperature recorder may be a useful addition. This will indicate precisely when a fault occurred and may help in assessing the extent of deterioration which has taken place.

If door closures become frozen, persons can be trapped in

the room and hence it is very necessary for an easily operated alarm or telephone to be installed inside the room. On no account should cryogenic materials such as liquid nitrogen or solid carbon dioxide be stored in the room, because they will rapidly displace the air and could cause asphyxiation. Flammable solvents in open vessels are extremely hazardous as the vapours form explosive mixtures with air and may be ignited by the thermostat contacts or other electrical equipment (see Chapter 7).

Hot-rooms

Microbiological work frequently requires the use of hot-rooms maintained at body temperature ($37°C$). Other work such as environmental studies may require hot-rooms, possibly at higher temperatures. Although there is an increasing tendency to provide individual incubators in the laboratories as an alternative to large hot-rooms, there are situations in which the provision of one or more such rooms is unavoidable.

The problems presented by hot-rooms are similar to those encountered with cold-rooms, although they are less likely to be used for storage. Since microbiology departments use these rooms for culturing organisms, there is a corresponding need for a very high standard of hygiene. As these organisms may be pathogenic, frequent disinfection is required and all materials must be considered as potential hazards. Surface finishes must be impervious and easily cleaned.

Most cultures are not destroyed by short-term falls in temperature, although their growth may be inhibited, but elevated temperatures caused by thermostat faults may well result in destruction of the organisms. Regular inspection and maintenance are therefore essential. It is good practice to install a secondary thermostat, set at about $1°C$ higher than the primary control system, which will operate in the event of a failure of the main control. It should, of course, be coupled to an indicator to show that a fault has occurred.

The comments made in reference to cold-rooms also apply in the main to hot-rooms, particularly in the matter of labelling materials and with regard to preventive maintenance. Unattended faults may not only result in unacceptable temperature variations but could also represent a considerable fire hazard. Flammable solvents should on no account be stored or used in hot-rooms as the high temperature results in increased vapour pressure. A solvent having a flash point of 30°C would probably be safe in the laboratory but in a hot-room at 37°C would be above its flash point.

Animal houses

In the UK, laboratory animals are the subject of several Acts of Parliament, the most important being the Cruelty to Animals Act 1876[8] (amended in 1973). Animals are used in laboratories for a wide variety of purposes and the Act applies only to vertebrates used for experimental work. For the purposes of the Act, an experiment is defined as a procedure, the outcome of which is not known in advance and in which the animals are being used to provide the answer to a question. Animals which are kept for such purposes as behaviour or breeding studies are not necessarily controlled by the Act, neither are those which are killed prior to being used.

It is important to realize that whether animals are classified as experimental or not, their housing and maintenance must be of the same high standard. Animals are capable of experiencing discomfort and distress as distinct from pain and it is the duty of all concerned to ensure that at all times they are properly housed, fed and maintained in clean and healthy accommodation[9]. Animals should only be obtained from reputable suppliers.

The subject of experimental laboratory animals is a very emotive one and a discussion on personal attitudes is not within the scope of this book. It is, however, important to realize that everyone has a right to hold particular views on the matter and to express those views either as an individual or as a member of a group. There is no doubt that animals have been used in unnecessarily large numbers or in ways which leave much to be desired, but for the purposes of this chapter we will confine our discussion to the problems associated with the management of animal houses operating within the framework of current legislation.

All mammals imported into the UK are subject to the

Rabies Act 1974, which was introduced to prevent the entry of rabies into the country[10]. The main requirement of the Act is that all imported mammals are to be kept in quarantine conditions for a period of six months. Administration of the Act is controlled by the Ministry of Agriculture, Fisheries and Food who appoint inspectors to ensure that correct procedures are adopted, that records are properly kept and that premises comply with the prescribed standards.

Animal house management problems are mainly associated with planning and allocation of accommodation, compliance with legislation, staffing, costing and safety.

In many biological laboratories, the animal house is a vital part of the organization and the work of the laboratories is totally dependent on the efficient supply and maintenance of a stock of animals. Planning of the unit is not without its problems and requires considerable thought at the early stages. Considerable demands are made on the mechanical services such as heating and ventilation and it may be virtually impossible to make any significant alterations should these prove to be inadequate[11,12].

Access is required for the very large quantities of food and cage litter needed and provision must be made for the disposal of the considerable amount of debris generated daily. This problem can be reduced substantially if the ground floor can be used, but if upper levels are utilized a large lift is essential.

The use of large rooms should be avoided; small rooms not only simplify the problems of housing a variety of species but reduce the risk of cross-contamination and the resultant spread of any infection in the stock. Successful animal breeding requires accommodation in a situation where noise and disturbance are reduced to the minimum possible levels, and animals used for work with infectious diseases must be housed in rooms which can be isolated from their surroundings. Animals in quarantine are required to be kept in rooms with double doors, only one of which can be opened at a given time. If specific-pathogen-free (SPF) animals are to be housed, it will be necessary to isolate rooms for this purpose. Large colonies of such animals will require a complete suite of rooms which will include ancillary rooms such as stores and cage-cleaning facilities, the entire suite being isolated from the rest of the animal house.

The heating system should be designed so that individual rooms can be maintained at different temperatures to suit the needs of particular species. It is good practice to use two independent systems, such as hot water radiators and electric heating panels, either of which automatically takes over from the other in the event of a failure. A high level of ventilation is essential, in the order of 10 air changes per hour, and the system must be so arranged that air extracted from one room cannot enter any other. Good lighting is required and, if nocturnal animals are to be housed, time clocks should be installed in the supply lines to permit a 'reverse daylight' system to be operated. Large windows will allow the maximum benefit from natural daylight but in rooms containing large animals the glass should be either reinforced or protected with wire mesh to prevent possible escape.

Surfaces will require frequent washing and must, therefore, be impervious and resistant to detergents and disinfectants. Ceilings should be finished in gloss paint and walls may be either tiled or gloss painted. Several impervious flooring materials are available, either trowelled finishes, such as asphalt, epoxy resins, etc., or quarry tiles laid on a cement base sub-floor. Materials based on asphalt are not entirely satisfactory as they are easily dented, and thus become difficult to disinfect. Floor drains must be fitted with grilles to prevent the waste pipes becoming blocked with debris. Large-bore pipes are essential throughout the drainage system, which must be independent of the rest of the waste system of the building and should be provided with cleaning-eyes at strategic points.

All fittings such as heating and lighting units must be designed and installed in such a way that they will not be damaged or become electrically unsafe when rooms are washed.

Apart from animal accommodation, the unit will require space for storage, cage cleaning and sterilizing, staff toilet and shower facilities and record-keeping. In some units it may be necessary to sterilize cage litter prior to disposal. Provision for incineration of litter and carcasses is required and this should be sited as near as possible to the animal house. Several store rooms are desirable, one for foodstuff, one for bedding and another for spare cages. Protective clothing used by animal house staff is not always acceptable to laundry companies and the provision of washing and drying machines is advantageous.

Cage-washing and sterilizing machinery is, of necessity, very large and provision for its installation must be made at the planning stage. Electrical or steam heating supplies must be available as required for this equipment.

Allocation of space for particular needs or to individual workers is frequently a source of friction and frustration. There is a tendency for requirements to change, sometimes drastically, and on occasions it may be necessary to reallocate rooms and to move animals to different accommodation to satisfy demands. Some workers consider that their prestige is proportional to the number of animal rooms that they use; in fact, the reverse is more often true and considerable diplomacy is called for if disputes are to be avoided. If, for example, a breeding colony is to be accommodated, the ideal situation would be in the quietest part of the unit, but if the only available room is adjacent to a noisy cage-washing area, some rearrangement of space allocation will be required. This must result in extra work for the staff and inconvenience for the worker whose animals are to be moved.

Mention has been made of the legal requirements associated with animal house management, particularly if the work is experimental[13]. Under the Cruelty to Animals Act, premises housing experimental animals must be approved by the Home Office, who appoint inspectors to ensure compliance with the Act and who may inspect the premises at any time. A licence is required by each person carrying out experiments, and for certain procedures certificates are also required. A licence on its own imposes a number of restrictions and basically requires that the animal must be anaesthetized throughout the procedure and must be killed before it recovers from the anaesthetic. Many of these restrictions may be lifted by obtaining the appropriate certificates, which are summarized in *Table 6.1*.

Licences and certificates are legal documents which are personal to the holder and delegation is expressly forbidden. A licensee may, however, permit an assistant to administer anaesthetics or to carry out mechanical duties such as holding an animal or administering a diet which he has prescribed.

Table 6.1

Procedure	Horses, asses, mules	Dogs or cats	Other vertebrates
Under anaesthesia without recovery	Licence + Cert. F	Licence	Licence
Under anaesthesia with recovery	Licence + Certs. A and F	Licence + Certs. B and EE	Licence + Cert. B
No anaesthesia used	Licence + Certs. A and F	Licence + Certs. A and E	Licence + Cert. A
Lectures and demonstrations under anaesthesia without recovery	Licence + Certs. C and F	Licence + Cert. C	Licence + Cert. C

Assistance of this nature is often provided by animal technicians and they do not require to be licensees if their involvement is limited to such tasks.

The inspectors also require all cages to be labelled with the name of the licence holder and a brief description of the purpose for which the animals are used. If they consider that animals are not being properly cared for, they may insist on changes being made, such as in the size of cages or the standard of hygiene. Their inspections also include records of work done and animals used, and failure to observe the requirements of the Act may result in the cancellation of registration of the premises or revocation of licences and certificates. Each licensee must also make annual returns detailing the number of animals used, including those used conjointly with other workers.

From the above it follows that accurate record-keeping is of major importance in animal house management. Not only is it required by the Act but a record of animal sickness or deaths is an indicator of the standard of management and may help to provide an early warning of problems associated with lowering standards or outbreaks of infection among the stock.

Records are also required to monitor such factors as breeding rates and consumption of food and bedding material, and an analysis of the records may indicate where improvements or economies could be made. Without such records, it is not possible to assess with any accuracy the cost of maintaining animals and the charges to be made to the departments concerned. It should be possible to calculate the cost of maintaining any species with considerable accuracy and these figures are also useful when forecasting budgets or making applications for funds.

The staff requirements will be related to the size of the animal house, the variety of animals used, the complexity of the work and the degree of involvement of the staff. In the past, animal technicians' duties were confined to the maintenance of the animals but in recent years there has been an increasing tendency, particularly in research establishments, to involve the animal house staff in the scientific work. It is rightly argued that such involvement develops more interest in the work and results in increased efficiency and enthusiasm. Training courses for animal technicians now place more emphasis on the scientific aspects of their work and cover such subjects as genetics, nutrition and animal diseases. Most animal technicians attend such courses on a part-time day-release basis and allowance must be made for their absence when assessing staffing levels. Their services are also required on a rota basis at weekends and on public holidays, and they may be given days off in lieu of this attendance or be paid for overtime working.

Regardless of the actual number of staff employed, it is essential to nominate one of them to act as supervisor and a deputy must be on hand to take charge in his absence. In a large unit, particularly where large animals are housed, much of the work will be of a manual nature and it is not unusual to employ several unqualified staff to carry out such tasks. The supervisor is responsible for day-to-day management of the work-force, including the organization of rotas and shift work, where applicable, and for the routine ordering of supplies. His experience and qualifications should enable him to assist with the in-post training of the junior staff and to handle most of the liaison work with the laboratories.

Accidents which occur in animal houses consist mainly of bites and scratches from animals, injuries caused by dropping heavy items such as cages, cuts caused by knives used in food preparation and burns from incinerators and sterilizers. Safety is an important part of animal technician training and first-aid facilities must be available. Many animals are known to carry diseases which can affect man and some of these can result in severe illness or even death[14]. Rabies has been mentioned, and such diseases as tuberculosis, Simian Herpes

(B) virus and Marburg disease have been known to infect animal handlers. Where possible, vaccination should be offered to staff if any such risk is thought to exist. If the presence of infections of a hazardous nature is suspected, the animals must be totally isolated and the problem investigated fully. Should the infection be confirmed, the appropriate authorities must be notified, the animals killed and the carcasses incinerated. It may, however, be necessary to retain certain organs for further investigation. Staff who may have been infected must be placed under medical supervision without delay.

Accommodation for animals must be designed and constructed in such a way that escape is impossible. If, for example, an animal were to escape from its cage it is essential that it is confined in the room. Doors should be self-closing to prevent their being left open inadvertently and the provision of windows in the doors enables handlers to make a visual check before opening rooms.

Equipment for handling heavy items such as cages and sacks not only reduces the physical labour but enables the work to be done safely and more efficiently. Machinery such as food mixers should be serviced regularly and staff must be instructed in their correct use.

Reprographic units

The ever-increasing volume of paperwork associated with the laboratory makes it necessary to provide a copying service for a wide range of documents of varying sizes such as letters, notices, charts, graphs, scientific and technical information.

Several systems are available for this work ranging from small and inexpensive desk-top copiers to large automatic units capable of a very high ouput. The choice of equipment will be governed by the expected workload and the maximum size of the originals. Until fairly recently, most copiers used photochemical systems but although some are still produced, they have been largely superseded by electrostatic copiers. The simpler units still require an intermediate negative to be made from which the final copy is produced, but the more sophisticated versions produce a finished copy in one operation. Many require a specially coated paper but some use uncoated paper. The latter type, known as plain paper copiers, are appreciably more expensive but if a large output is required the cost per copy is lower — an important consideration when deciding which equipment to buy.

Modern copiers are very rapid in use, need little skill and are capable of producing a predetermined number of copies from each original. They are relatively inexpensive and it is quite feasible to have a number of them in use, either in one room or distributed at strategic points in the building, e.g. in the offices and in the library. A number of manufacturers make their machines available on lease which may have certain cost advantages, bearing in mind that charges also include servicing contracts. Paper is supplied either in precut sheets of sizes to suit the equipment or as a roll which is guillotined automatically during the copying process to the size set on the controls. The latter method reduces wastage to the minimum.

It is usually more satisfactory to make one person responsible for the service and to record the number of copies made. The ordering of materials is better controlled and charges can be accurately assessed. In small organizations there may be advantages to providing an uncharged service and allowing staff to use the copiers without having to keep records. This method completely removes administrative costs, which could exceed the cost of the actual copies.

Where copies of typescript are required, the most efficient method is to use a stencil duplicator. The stencils are cut on a typewriter and with a good duplicator will produce several hundred copies. Coloured inks can be used and paper is also available in a wide range of colours. Duplicators are made for either hand or electric use, the latter usually having provision for a predetermined number of copies to be made. The cost per copy is substantially lower than with electrostatic systems but the cost of producing the stencils must be taken into account. Electronic stencil-cutting equipment can be used for reproducing diagrams and photographic work, but as the equipment is costly it is usual to arrange for such work to be done by specialist firms.

The production of large numbers of copies of items such as booklets, staff instructions and similar publications is normally carried out by conventional printing processes and requires expensive equipment and trained staff. It is not usual, therefore, to attempt such work within the organization but to contract out to printing companies.

Document copiers are unfortunately open to a certain amount of abuse; it is not unheard of, for example, for them to be used for copying articles from non-professional sources such as books, magazines, knitting patterns and so on. In the UK, most published material is covered by legislation relating to copyright; reproduction for certain purposes is therefore expressly prohibited.

Laboratory workshops

An area set aside for the purpose of carrying out minor modifications, repairs and adjustments to apparatus is of considerable benefit even to the smallest of laboratories. In a large organization there is a very definite need for a well-equipped and adequately staffed workshop, or, indeed, several workshops, to provide the facilities and expertise to keep apparatus in good working order and to design and construct prototype equipment. This is particularly important in teaching establishments where apparatus is used in large quantities and in research organizations where special items of equipment are needed for original work.

The space, staff and range of tools and machinery required are related to the size of the organization and the variety of work carried out and must be considered in the early planning stages. As workshop machinery is somewhat heavier than laboratory apparatus and tends to generate considerable noise and vibration, it has been common practice to site workshops in basements. There is little justification for this and the ground floor should be chosen where possible. Floor area should be spacious, because there is frequently a need to handle very sizeable sheets of material and bulky items of equipment. It is advantageous to divide the area into a number of separate rooms, each used for a specific purpose such as heavy machining, delicate assembly and electronic work. A clean subsection should be set aside for the preparation of drawings and for the paperwork associated with running the workshop. Benching must be of very substantial construction, standard laboratory benching being generally unsuitable for workshop use.

A high level of illumination is required[15], which must be reinforced with local lighting over machines and where detailed assembly or adjustment is involved. Fluorescent lighting is normally used to provide the general illumination but it has the disadvantage of making rotating machinery appear to be stationary at speeds which are related to the frequency of the mains supply. This problem can be overcome by connecting adjacent lighting units to different phases of the supply or by using tungsten lamps over machines.

A high level of ventilation is essential[16], particularly if welding or paint spraying are carried out. It is generally preferable for the workshop ventilation system to be independent of the remainder of the building and to exhaust directly into the atmosphere by the shortest possible route. Electrical supply demand is usually higher than in the laboratories and there are many advantages in having a three-phase supply available. Motors used for driving machines are smaller and more efficient than their single-phase counterparts, and their lower current ratings require less heavy cables. The supply should be fitted with an isolator, which can be operated from several parts of the workshop, to cut off the power in an emergency. A gas supply should also be provided and is useful for many purposes such as the heat treatment of metals, and small soldering and brazing work. Hot and cold water supplies are also essential and the provision of a compressed-air supply may be an asset.

Storage facilities for materials and components should be provided in the form of a room separated from but adjacent to the main workshop. Stock material is likely to be extremely heavy, so racks and shelving must be of adequate proportions to carry the loads. Many materials are supplied in lengths of up to 5 m; allowance for this fact must be made when considering delivery access and the movement of stock from the store to the workshop.

The selection of machinery and tools is best made in consultation with the technician responsible for running the workshop, bearing in mind the variety of procedures to be used. These will almost certainly include turning, milling, grinding, boring, brazing, welding and sheet metal work. A wide range of materials are used in laboratory equipment; apart from the commonly encountered metals, a variety of plastics, wood, etc., are employed and the workshop must be able to work in most, if not all, of these. The capacity of the machines must, of course, be sufficient to handle the largest work anticipated and it may be necessary to provide more than one of each kind. For example, a large lathe is required for heavy turning and an instrument lathe is necessary for delicate work. An increasing amount of electrical and electronic apparatus is finding its way into laboratories and facilities should be available for their design, construction and repair.

The number of persons employed in the workshop and the degree of skill required of them can only be decided in relation to the volume and range of work, but it is essential that one person should have overall responsibility. He should obviously have a wide range of skills, but is unlikely to be an expert in every field. It may, therefore, be necessary to appoint one or more specialists, such as one who is experienced in electronics. The training received by workshop technicians is very different from that of laboratory technicians and it is sound policy to encourage the interchange of ideas between the two groups, thus enabling each to complement the work of the other.

The workshop can make a considerable contribution to the economy of the organization but it should not be used for the production of items which are readily obtainable from trade sources unless these can be made at an appreciably lower cost or to a higher standard than those normally available.

Requests for work to be done should be made in writing or on a form prepared for this purpose. They should be accompanied by drawings, sketches or circuit diagrams, etc., as appropriate. To avoid misunderstandings regarding what is actually needed, it is generally more satisfactory if the person making the request discusses his precise requirements with the technician who is to do the work. This is particularly important where there is a choice of materials which could be used or if the design could be improved or simplified with slight modification. Scientific staff are not always conversant with engineering methods and frequently ask for very complex equipment to be made, unaware that a simpler design would work just as well, if not better.

In a busy workshop, it is seldom possible for any but the

most urgent jobs to be done immediately and the expected completion date should be advised. Delays may be caused by materials not being in stock, machines not being available or the fact that other work is outstanding. Conflict may arise when several people are waiting for work, in which case it may be necessary to allocate priorities for each job and the laboratory manager may be called upon to determine the order in which the work should be done.

Costing is usually carried out on a materials only basis or on a time and materials basis. In order to assess the charges, records must be kept of the materials and components used and of the time spent on each job. This information can be recorded on the original request form as the job proceeds and then accounts can be presented to the various departments at regular intervals.

Drawings or circuit diagrams relating to work done should be retained for reference in case they are needed at a later date. Any servicing data and diagrams supplied by equipment or component manufacturers should also be filed for future use.

Frequent demands are made on workshop technicians to carry out private work for other members of the staff and a few words on this subject would seem appropriate. It is difficult to raise any valid objection to the occasional request for a small repair job to be done, but in a large organization it is clearly not feasible to extend this privilege without restraint. If such a situation were allowed to develop unchecked, the workshop would have insufficient time to do its official work and could consume a vast amount of material. A great deal of diplomacy may be required on the part of the technician in charge and he may find it less embarrassing to refer all such requests to the laboratory manager who is usually in a more advantageous position to refuse the persistent scrounger. Requests of this nature are often made by persons who are unwilling to pay for these jobs to be done through the normal channels and who take advantage of the situation.

Similar problems are associated with employees wishing to use the workshop facilities or to borrow tools for their own use. The technician will need considerable tact to dissuade those who, in his opinion, lack the necessary skill and may be a danger to themselves and others, or who may damage the workshop equipment. It is sometimes difficult to believe that so many people do not possess such simple tools as a hammer or screwdriver.

Workshops are potentially dangerous areas and safety is therefore an important topic, much of which is discussed in Chapter 7. New legislation applicable to workshops is published at intervals and staff should be made aware of all such regulations and the need to observe them[17-23]. Notices are available summarizing the various requirements and these should be displayed in the workshop. Examples include the Protection of Eyes Regulations and the Woodworking Machines Regulations. If at all possible employees should not be permitted to work alone in a machine shop. Guards are required on some machines and these must be checked regularly to ensure that they have not been removed, tampered with or modified. All equipment should be turned off after work, especially cylinders of gas such as acetylene.

Cleanliness is as important in the workshop as it is in the laboratory. Not only does it provide a more efficient working environment but dirty floors may be slippery, particularly if oil or grease is allowed to accumulate. A bench cluttered with partly dismantled electrical equipment is an obvious hazard, whilst rags and other rubbish dumped in corners represents a serious risk in the event of a fire. Commonly occurring workshop accidents include cuts from tools and burred metal edges, burns and injuries to the eyes, damage to the hands from machinery and injuries caused by heavy objects being dropped. First-aid facilities must be available in or near the workshop and staff should be adequately trained in accident prevention.

The Health and Safety at Work etc. Act 1974 requires that all equipment made for use at work must be safe in normal use and this requirement applies, of course, to equipment made in the workshop for use in the laboratories. It also applies to repair and maintenance work on equipment and it is, therefore, essential to ensure that any item made or repaired is safe to use before it leaves the workshop.

In the interests of the safety of the workshop staff, no equipment in a dangerous condition should be sent for repair unless it is accompanied by a clear warning of the nature of the hazard. For example, a centrifuge contaminated with a dangerous chemical must be thoroughly cleaned before being moved to the workshop; a water bath which is electrically live must carry a warning notice to this effect.

Some materials used in workshops have a considerable value as scrap and whilst many off-cuts can be used for making small items, it is sound economic policy to provide bins for scrap copper, brass, aluminium, lead, etc., so that their contents can be sold to scrap dealers at intervals. Similarly, metals recovered from obsolete apparatus can be treated in the same way. The income raised could be used, for example, to finance the workshop petty cash account or other small budget. It is doubtful whether it is worth saving small electronic components recovered from old equipment but items such as electric motors are useful for building prototypes or one-off instruments required for a limited period.

Audio-visual aids

The use of audio-visual aids in laboratories has increased considerably in recent years and there is every reason to believe that this trend will continue. In the educational field, in particular, the separate audio-visual aids unit has become a feature of all but the most modest establishments. This class of equipment plays a vital role in research departments as an aid to investigation and presentation of results. Lectures and demonstrations are almost invariably supported by slides or films.

A wide range of audio-visual equipment is available today and is able to cater for almost every demand.

AUDIO AIDS

For lecture use, a microphone, amplifier and loudspeaker combination has been considered satisfactory for many years but recently radio microphones have become available and prices are now sufficiently competitive for these to be worthy of serious consideration. They have the great advantage of enabling the lecturer to move about whilst speaking. Where several speakers are involved, such as in a discussion, the one receiver is capable of handling any number of transmitter units.

Tape recorders have numerous applications, such as recording observations in the laboratory, discussions, noise generated by machinery, animal sounds and so on. Highly sensitive units may be used to record very low levels of sound which could not be detected by ear. The reel-to-reel machines are ideal for studio use but the quality of portable cassette recorders has improved to such an extent in recent years that they are perfectly satisfactory for all but the most exacting work. When playback is effected through a high-quality amplifier and speaker system, the reproduction is excellent. Their ability to operate from batteries and the low cost of cassettes are obvious advantages. Copies can be made easily and cheaply and the cassettes are small enough for an efficient filing system to be operated.

VISUAL AIDS

The 35 mm slide in a 2 in by 2 in mount is accepted as a universal size. Slides may be made directly from reversal film and for charts, diagrams and line drawings, positives are made by photocopying black-and-white negatives on to black-and-white film stock or diazo materials if coloured slides are required. Instant picture systems may also be used for producing black-and-white slides directly from originals. Projectors may be fitted with manual or automatic slide changers. The latter may be controlled by signals on magnetic tape cassettes to provide slide changes at the appropriate points in recorded lectures. The important features of any slide projector are the light output, which must be sufficient to produce a bright image of the chosen size; the heat output, which should be low enough not to damage the slides; and the focal length of the lens. The latter must be chosen to suit the size of the room in which it is to be used and it may, therefore, be necessary to stock more than one lens. (The magnification is approximately equal to D/F, where D is the distance between projector and screen and F is the focal length of the lens.)

Some projectors have facilities for projecting 35 mm film strips or for projecting directly from microscope slides.

Adaptors are also available to enable microscopes to be used for projection but these are only capable of producing small images up to about 30 cm square and are, therefore, not suitable for large audiences.

Overhead projectors are extremely useful in the lecture room. The material for projection is prepared on transparent sheets or roll which, after use, can be cleaned and re-used. As the transparencies are fairly large (about 20 cm square) a high degree of magnification is not necessary and the equipment is somewhat cheaper than slide projectors of equivalent performance.

Cine films have considerable visual impact and are, therefore, of great value in teaching and industrial training. In research, they are frequently the only way in which events can be recorded and studied, especially if the subject is fast moving and can be filmed in slow motion.

The film size in most common use is 16 mm, and this is available in sound or silent versions. The sound may be recorded as an optical or magnetic track. Super 8 mm is now made to a very high standard and with suitable equipment is capable of producing high quality films. Sound may be added to a magnetic strip on the film or from a separate tape recorder synchronized with the projector. Many films are available on 35 mm stock, but as both film and equipment are expensive and bulky, this size is not commonly used in the laboratory. Cine films may be made as part of the work or purchased from specialist firms. Alternatively, they may be hired if required for only a short period. A range of useful and interesting films may be obtained on such subjects as safety, design or on specialized scientific and technical matters.

Time-lapse cine is a method of filming events which take place over a relatively long period. The equipment is set to photograph individual frames at regular time intervals, say one per minute. When the film is processed and projected at normal speed, an event which may have taken days or even weeks to occur may be viewed in minutes. The equipment may be coupled to a microscope to observe minute changes of shape or structure.

The choice of a cine projector is in many ways similar to the selection of a slide projector, the focal length of the lens again being an important point. The projector must also be capable of accommodating the largest size reel which is to be used, and of reproducing sound from magnetic or optical track or, in some cases, both types.

An adaptation of the cine projector is the cine analyser, which is basically a projector in which the film can be advanced one frame at a time. The image is projected on to a ground-glass screen fitted with vernier scales which enable precise measurements to be made. Episcopes are used to project images from opaque materials such as books, leaflets and photographic prints.

The projection of both slides and cine films requires the use of screens and these may be either a permanent feature of the rooms used or portable screens which may be used anywhere. Back-projection is a useful alternative where space is restricted. The image is projected on to the back of a translucent screen which is viewed from the front. Projectors with this facility built in are available and some accept the film in the form of a cassette, which is simply inserted in the instrument.

Closed circuit television (CCTV) is an extremely useful aid to observation, particularly in situations which cannot be observed by eye. Examples include studies relating to explosives and impacts and in high temperature or radioactive environments. The system is also invaluable where a large audience requires to study a small-scale event such as surgery. CCTV will also enable students to observe the techniques from outside the operating theatre. The camera may be hand controlled or operated remotely as would be the case in a hazardous situation. Any number of monitors may be fed from the one camera outfit. Black-and-white or colour systems are available and the results may be recorded on video tape or cine film for further use. The tape or film may also be edited to remove unwanted material or to rearrange sequences.

The scope of closed circuit television systems may be extended by the use of video tape recording equipment. Recorders are available at fairly modest prices and can record in black-and-white or colour from television cameras or from broadcast programmes. Some machines use tape cassettes, whilst the more sophisticated models operate reel-to-reel. For scientific work, the most versatile equipment may be used in slow motion or 'stop action' modes. Picture quality is not as high as that obtained with cine film but, as no processing is required, instantaneous viewing is possible. The tape can, of course, be erased and used again if a permanent record is not required.

Equipment for editing and splicing film and tape is not expensive and should be available in the unit.

The organization of an audio-visual aids unit consists primarily of appointing a member of staff to run the unit to a well-designed system. A store room is required for the equipment and, in view of its attractive nature, this room should be secure against theft. A stock record must be kept of all the equipment held and some system of booking-out materials and apparatus must be established. If demand is likely to be high, bookings should be made well in advance, as should reservation of the rooms required for projection. It is unwise to loan equipment for personal use, although it may be issued to other departments on a temporary basis provided such transactions are recorded.

The technician in charge should also be responsible for maintenance and repair of the equipment, the hiring of films and their return after use. As he would probably do much of the projection himself, he will be aware of the condition of the equipment and should be capable of carrying out routine maintenance and minor adjustments. The records should also make it possible to cost the various activities and charge them to the departments concerned.

In the UK, advice on audio-visual aids may be obtained from the National Audio Visual Aids Centre, London.

Glassblowing shops

Any organization using large quantities of glassware will probably benefit from having one or more glassblowers on the staff. Not only can repairs and modifications be carried out quickly and cheaply but special items not generally available can be made to order.

The size of the room will be related to the number of persons employed and each will require a work bench together with the appropriate burners and tools. If large items are involved, it will probably be necessary to install one or more glass-working lathes. An annealing oven, capable of accommodating the largest item to be made, must also be provided. Good lighting is essential and a high level of ventilation is required to compensate for the considerable amount of heat generated in the room. Oxygen and compressed air supplies are necessary as, of course, is a gas supply. This would normally be town gas but in areas where this is not available, bottled gas may be used as an alternative. All gas supplies must be fitted with non-return valves to prevent any possible blowback of air or oxygen into the gas line. If quartz is to be worked, a supply of hydrogen must also be available.

A stock of tubing and rod will be required, together with components such as standard joints, and a suitable area must be provided for their safe storage.

All requests for work should be made in writing and accompanied by drawings where appropriate. Any item for repair or modification should only be accepted if it is clean and dry, and every effort should be made to ensure that it is free from hazardous contamination. A more satisfactory job usually results from discussion with the glassblower who, as a result of his training and skill, is probably the best person to advise on design.

Charging of work is normally on a time and materials basis, accounts being submitted to the departments at regular intervals.

Although accidents are not uncommon in glassblowing shops, most are of a fairly minor nature due mainly to the fact that glassblowers are familiar with the properties of glass and have a healthy respect for the dangers it presents. The most common injuries are cuts and burns, for which first-aid materials must be readily available. The presence of gas cylinders in the room is an obvious hazard and these must be treated with caution (see Chapter 7). Cases have been reported of cylinders exploding for various reasons such as overheating. Scrap glass must be deposited in bins provided specifically for that purpose and not mixed with other rubbish such as packing materials. Protective clothing must be provided, including goggles and heat-resisting gloves. All equipment must be securely shut down at the end of each work period, except those items which are specifically required to be left running. Until recently, many glassblowers had little or no formal training and the traditional apprentice-type approach has been the way in which these most useful craftsmen have acquired their expertise.

REFERENCES

1. Breach, M.R., *Sterilisation, Methods and Control.* Butterworths, London (1968)
2. The Radioactive Substances Act 1960. HMSO, London
3. The Ionising Radiations (Unsealed Radioactive Substances) Regulations 1968. HMSO, London
4. *Code of Practice for the Protection of Persons Exposed to Ionising Radiations in Research and Teaching.* HMSO, London (1968)
5. Hughes, D. and Cullingworth, R., *The Design of Laboratories for Radioactive and other Ionic Substances.* Koch-Light Laboratories Ltd, Colnbrook, Buck (1971)
6. *Monitoring of Radioactive Contamination on Surfaces,* IAEA Technical Report No. 120. HMSO, London (1970)
7. The Copyright Act 1956. HMSO, London
8. The Cruelty to Animals Act 1876. HMSO, London
9. *The UFAW Handbook on the Care and Maintenance of Laboratory Animals.* Churchill Livingstone, London (1976)
10. The Rabies Act 1974. HMSO, London
11. Clough, G. and Gamble, M.R., *Laboratory Animal Houses. A Guide to the Design and Planning of Animal Facilities.* Medical Research Council Laboratory Animal Centre, Carshalton, Surrey (1976)
12. *Notes on Animal Houses.* HMSO, London (1971)
13. North, P.M., *Modern Law of Animals.* Butterworths, London (1972)
14. Fiennes, R.N.T-W., *Zoonoses of Primates.* Weidenfeld, London (1967)
15. *Lighting in Offices, Shops and Railways Premises.* Booklet No. 39, Health and Safety Executive; HMSO, London (1976)
16. *The Ventilation of Buildings, Fresh Air Requirements,* Technical Data Note No. 19, Health and Safety Executive; HMSO, London (1976)
17. The Health and Safety at Work etc. Act 1974. HMSO, London
18. The Factories Act 1961. HMSO, London
19. The Prescribed Dangerous Machines Order 1964. HMSO, London
20. The Asbestos Regulations 1969. HMSO, London
21. The Abrasive Wheels Regulations 1970. HMSO, London
22. The Protection of Eyes Regulations 1972. HMSO, London
23. The Woodworking Machine Regulations 1974. HMSO, London

FURTHER READING

Coates, M.E. (Ed.), *The Germ-free Animal in Research.* Academic Press, London and New York (1968)
Ionizing Radiations (Sealed Sources) Regulations 1969. HMSO, London
Safety of Quick Opening and Other Doors of Autoclaves, Technical Data Note No. 46 HM Factory Inspectorate, London (1974)
Short, D.J. and Woodnott, D.P. (Eds.), *The I.A.T. Manual of Laboratory Practice and Techniques.* Institute of Animal Technology, London (1969)
Taylor, D.M. and Taylor, M.P., 'Radiation hazards in the laboratory', *Laboratory Equipment Digest,* 5(2), (1967)
The Radiological Protection Act 1970. HMSO, London

7
Health and Safety

The basic approach

Many countries have introduced legislation which deals with the health and safety of various groups of work-people. Whilst the USA and several European countries (particularly Scandinavia) have now instituted occupational health Acts, there is no doubt that one of the most comprehensive pieces of safety legislation is the United Kingdom's Health and Safety at Work etc. Act 1974[1].

This Act was built on the report of the Robens Committee and was implemented on 31 July, 1974[2]. Prior to that date, there had been many attempts to legislate for industrial and environmental safety, but most of this legislation was on a piecemeal basis. The Factories Acts 1937 and 1961[3] were probably the most widely encountered of these Acts and although much was achieved by their application they always lacked really effective powers of inspection and enforcement.

Other Acts dealt with safety in specified places, such as mines, quarries and tips, whilst still more Acts were concerned with safety in the use of specific substances, such as alkalis, radioactive substances[4], and petroleum spirit[5], etc. Much of this previous legislation was of no direct interest to the laboratory manager, but the Health and Safety at Work Act has become of major importance in the performance of his daily task.

THE HEALTH AND SAFETY AT WORK ETC. ACT 1974

This is an all-embracing enabling Act, uncluttered with technical detail. It covers all persons at work including those working in educational establishments and research laboratories. The Act is related to persons at work rather than the substances used or the tasks performed. It has been estimated that some 5−7 million persons in the UK who were not previously covered by the Factories Acts and the Offices, Shops and Railways Premises Act 1974[6] are now covered by specific safety legislation. It also provides protection for persons who, although they may themselves not be at work, are likely to be affected by the activities of those who are. It places wide-ranging duties on every employer to 'ensure, so far as it reasonably practicable, the health, safety and welfare at work of all his employees'. Similar duties are placed on persons in control of premises, persons who design, manufacture, import or supply any article for use at work, or who manufacture, import or supply any substance for use at work. Duties are also placed on employees to take care of themselves and others who may be affected by their actions, and to co-operate with employers to ensure safe working conditions.

The Act is administered by the Health and Safety Commission, which is able to recommend to the Secretary of State that regulations be made under certain sections. The Commission also has responsibility for the promotion of research, making provision for training and for dissemination of information. It is also able to issue, or to approve, codes of practice to supplement the regulations. The enforcement of the Act is the responsibility of the Health and Safety Executive which, as well as exercising its new powers, has taken over many of the powers previously held by the Factory Inspectorate[3]. The Executive appoints regional and local Inspectors whose task is to inspect workplaces of all types and to investigate accidents and dangerous occurrences, to prosecute offenders and to issue Improvement or Prohibition Notices.

Thus laboratory managers, chief technicians, etc., as an integral part of line management, are now directly and vitally concerned with the legal aspects of laboratory safety.

Organization of laboratory safety

The arrangements for laboratory safety in a given establishment may be chosen from one of several alternatives as appropriate to the size of the establishment, the number of persons employed and the nature and range of hazards encountered.

LINE MANAGEMENT

The employer carries the prime responsibility for the safety of his work-force and he may choose to delegate certain powers down the chain of command via departmental heads, section leaders, chief and senior technicians, etc., but the ultimate responsibility for safety remains with the employing authority, managing director, principal, etc. Clearly the employer would find it impossible to supervise all the day-to-day arrangements in a large company, and he will need to rely on line management for much information and advice.

SAFETY OFFICERS

These are employees appointed by management to supervise and monitor the safety arrangements in a given field or area of work. For example, one often encounters radiation safety officers who are generally responsible for the supervision of radiation safety in a particular establishment or large department. It is essential that the powers, duties and responsibilities of safety officers are clearly and precisely defined. It is important that if in the event of an accident he is likely to be held responsible, then he should have the necessary powers to make and enforce the rules by the application of which the accident could have been avoided. Clearly, responsibility goes hand in hand with authority and ability to make and enforce the rules. The safety officer's prime duty is to keep a watchful eye on the day-to-day activities in his area or field of responsibility, but he should also have time at his disposal to keep abreast of developments in his field and be able to develop new safety arrangements to meet the changing circumstances of his establishment. Safety officers need to make periodic inspections of laboratories and workshops to ensure that the safety rules and regulations are being complied with. Generally, he will report to higher management, although he will be expected to give advice at all levels. To fulfil these task adequately, safety officers should be persons of wide experience with tact and understanding and with considerable powers of leadership.

SAFETY COMMITTEES

In many larger establishments it is necessary to set up safety committees. The Health and Safety at Work Act makes provision for these, and draft regulations which give wide powers to recognized trade unions have been published. Each recognized trade union in a workplace will have powers to appoint safety representatives. When two or more safety representatives have been appointed, it is envisaged that they will be able to request the employer to set up a safety committee and this the employer is bound to do within three months.

Many establishments in the industrial, educational and research fields will previously have inaugurated safety committees. Such existing committees will generally have been set up as part of a management exercise, in that members will have been selected or appointed by the employer to support the normal chain of command. To comply with the Health and Safety at Work Act, it will be necessary to reorganize these existing committees in order to take account of the new powers given to staff safety representatives. It will be essential to ensure that a reasonable balance is achieved between management and staff representatives and that a suitable range of expert knowledge is available to the committee. Safety committees should have clear terms of reference which specify the number of members and how they are to be appointed. The composition and strength of the committee will be related to the size of the establishment, the number of staff employed and the range of hazards likely to be encountered. The terms of reference should also state how often the committee will meet and define its powers of inspection. There should be provision for the investigation of accidents and near-miss situations and arrangements should be made for the publication of minutes and reports. The effectiveness of the committee will depend on, and will be judged by, the workforce; opinions will very largely be conditioned by the nature and quality of its reports and on the actions taken as a result of its deliberations.

CODES OF PRACTICE

Apart from the codes of practice approved by the Health and Safety Commission, it is likely that various trade associations, universities, colleges and industrial groups, will produce (if they have not already done so) specific codes of practice suitable for their particular circumstances. These codes of practice will give guidance to management and staff concerning the methods of working that are generally considered to be safe and reasonably practicable.

The purpose of these codes is to outline safe methods of working and to specify the most favourable conditions under which the work may be carried out, thereby reducing the accident rate in a particular industry or establishment. The codes have no legal standing, i.e. one cannot be prosecuted for failure to comply, but no doubt a court would take into account failure to comply with a reasonable and generally accepted standard when deciding the level of damages or compensation in a civil action. Codes of practice should be short, concise and easily remembered. A copy should be issued to each employee for his personal use and reference.

GENERAL ATTITUDE TO LABORATORY SAFETY

Fortunately, most laboratory accidents are of a relatively trivial nature, resulting in no permanent injury to the persons involved, and the cost of repairing or replacing the damaged equipment is relatively slight. Inspection of the accident records of almost any laboratory will show that the most common form of laboratory accident is laceration of the hands, most of these being caused by the careless use or manipulation of laboratory glassware. For example, junior (and occasionally quite senior) staff cut their hands when attempting to push unlubricated glass tubing through holes in rubber bungs. Serious accidents occur much less frequently, but in recent years a number of accidents have occurred where the failure to observe fairly simple rules or to take certain basic precautions has led not only to loss of life but very high financial loss to the employer concerned. Accident books usually contain records of personal injury and rarely reflect the extent or cost of the damage to property, nor do they show the number of occasions when a near-miss situation arose but in which no personal injuries were sustained.

It is important that each laboratory should inaugurate a set of basic rules for the prevention of accidents and that each member of staff is aware of the hazards likely to be encountered in his particular field of work. It is essential that all new entrants, particularly young persons, are guided and encouraged by their more senior colleagues to adopt safe working practices. Many accidents could be avoided by good housekeeping. Laboratories should be clean and tidy, unused equipment should not be left lying about and corridors or gangways should not be obstructed. All combustible packing and rubbish must be removed to a safe place as soon as possible. An untidy laboratory is a direct encouragement to staff to become slack in their general attitude to safety. It is important that senior staff should at all times set a good example to others and it is a function of management to ensure that staff at all levels behave in a responsible manner.

Occasionally, one meets a person who seems to be rather more accident prone than most of his colleagues; these so-called accidents may be caused by lack of concentration or, less often, by an almost deliberate lack of care or concern for others. Lack of concentration in an individual may be due to worry about some particular problem and if the pressure can be lessened or removed his performance may improve. So far as the person who exhibits little concern for the safety of his colleagues is concerned, it may be necessary to suggest that he finds employment in an occupation where he will be less likely to become the cause of a serious accident.

ACCIDENT BOOKS AND RECORDS

Many countries have implemented legislation that requires the maintenance of a record of accidents to persons[3,6]. In the UK, a special preprinted booklet may be obtained from HMSO wherein may be entered brief details of the accident and the injuries sustained. The place, time and cause, together with names of any witnesses, are also recorded. Apart from the legal requirement to keep accurate records, these books are useful because, if the type and frequency of accidents are known, it is possible to detect trends and to take the appropriate preventive action and thus reduce the overall rate of occurrence. A statement written soon after the accident may become useful to both the employee and employer in any subsequent litigation.

NOTIFIABLE ACCIDENTS

Apart from a duty to record in the official accident book all accidents involving personal injury, there is also a duty to inform the appropriate government department of a 'notifiable accident'. In the UK, employers are required by the Factories Acts[3] to notify the District Factory Inspectorate of any accident or disease which causes death, or disablement of more than three days' duration. The report is made on a prescribed form. In the case of serious injury or death, an inspector will almost certainly make an investigation and a prosecution may follow. This legislation is likely to be subsumed by the provisions of the Health and Safety at Work Act. In future, it is anticipated that the investigation will be carried out by the district or regional Health and Safety Inspector of the Health and Safety Inspectorate.

The hazards

FIRE

The annual cost of the losses due to fire both for industrial and educational establishments has been rising steadily for some time[7]. Several fires have occurred recently in the chemical industry in which the replacement cost of buildings, plant and equipment, plus the cost of dislocation and loss of production, has been very considerable. The universities have not escaped from the enormous increases in the cost of serious fires. There have been a number of fires in university buildings where, although the cost of the immediate physical damage to the fabric of the building was fairly low, the resultant damage by smoke, or from the corrosive products of combustion, to complex electronic equipment has been very high.

Many large fires arise from small beginnings; electrical equipment inadvertently left on overnight, failure of the water supply in a cooling system, self-ignition of packing materials, lighted cigarettes carelessly left on bench tops, and so on. It is clear, therefore, that if this annual toll is to be reduced, laboratory managers will need to develop an increasing awareness not only of the hazards but also of the precautions that must be observed. Every establishment should have a plan or code of practice for the prevention of fire, the secondary aims of which should be to increase the chance of successful containment of a fire once it has started so that the resultant damage may be reduced[8].

FIRE PREVENTION

Much could be done to reduce the risk of fire simply by improving the standard of housekeeping in laboratories. It must be impressed on all staff that persistent neglect of simple precautions will almost certainly lead to disaster. There should be provision for the daily removal of rubbish of all kinds. Combustible packing material must be removed to a safe place (preferably outside the main building) without delay and not stored in the laboratory for possible future use. Quantities of inflammable solvents should be kept to a minimum. Those which are required for ready use should be kept in approved storage conditions, whilst the remainder should be kept in properly constructed solvent stores. Staff using inflammable solvents should be aware of the significance of their flash points, auto-ignition temperatures and explosive limits in air, together with their toxic properties; a knowledge of available alternatives is also desirable. Laboratory staff should be encouraged to use less hazardous and preferably non-inflammable solvents whenever possible. Some solvents, such as carbon disulphide, have such low ignition temperatures that they may be ignited by exposure to hot steam pipes, whilst other solvents have such high vapour density that a layer of explosive vapour may form at floor level or in the troughs between benches. If high vapour pressure is combined with low flash point, as in diethylether, then the use of such solvents may be particularly hazardous. It must further be mentioned that a number of solvents form particularly unpleasant products on pyrolysis.

Electrical equipment and wiring must be properly maintained; in particular, worn flexible leads and damaged plug tops are a common source of trouble. The use of multi-socket adaptors should be discouraged. Many fires, both laboratory and domestic, have their origins in the use of unsafe electrical equipment or overloaded wiring. Smoking may be allowed in laboratory offices, libraries, etc., but it should be discouraged (if not forbidden) in any laboratory, because apart from the risk of fire there is also a further risk of poisoning or infection in laboratories where chemicals or micro-organisms are used.

Fire detectors and alarms

Many laboratories are now fitted with devices that are sensitive to smoke, sudden rise in temperature or increase in the amount of ionized gas in the air. When activated, they may cause a local or general alarm to sound or may automatically summon the fire brigade; they may also indicate the location of the fire at a central control point and operate fire-fighting equipment. Automatic sprinklers are not generally recommended for laboratories because of the damage likely to be caused by spurious alarms. It is important that all staff are familiar with the sound of the alarm bell and are aware of the action to be taken when the alarm sounds. The alarm must be audible in every part of the building including enclosed or isolated rooms such as photographic dark rooms, cold-rooms, toilets, etc.

FIRE-FIGHTING EQUIPMENT

Suitable and adequate fire-fighting equipment must be provided so that fires, once started, are detected, contained and extinguished without delay.

Water is one of the most effective and commonly used

extinguishing agents. Not only does it blanket the flames and deprive the fire of its supporting oxygen but it also provides cooling, thus helping to prevent re-ignition with the resulting spread of the fire to surrounding areas. Water should not be used on electrical fires or on fires involving burning oil or solvents immiscible with water. Large diameter hoses are best left to the fire brigade, because much damage may be caused by their unskilled use. Nevertheless, hose reels of less than 25 mm diameter and gas-propelled water extinguishers are widely installed for use against fires involving wood, paper and similar materials.

Foam extinguishers

Various types of extinguisher are used to blanket oil and some solvent fires, but the corrosive nature of some of the foams may cause considerable damage.

Carbon dioxide extinguishers

These extinguishers are excellent for many types of laboratory fire and, if used promptly, they are very effective. The noise emitted when they are used may act as a warning to other staff. They leave no mess or residue but need to be used with some care on bench fires, since bottles, flasks and other lightweight containers may be overturned by the rapidly ejected stream of carbon dioxide. They must not be used on fires involving burning alkali metals.

Dry powder extinguishers

These extinguishers are filled with a relatively inert, finely divided powder, such as magnesium oxide or sodium bicarbonate, which may be rapidly ejected from the container by means of carbon dioxide or compressed nitrogen. The small disposable cylinders are of very limited use because of the very small charge which they contain, but the large cylinders containing upwards of 2 kg of powder are effective against most types of fire with the exception of burning metals.

They create a fair amount of mess which is generally easily cleared up. They should not be used on fires involving complex or delicate mechanical equipment unless it is absolutely unavoidable.

Carbon tetrachloride extinguishers

Such extinguishers are now very rarely used because of the toxic nature of the filling, together with the products of its pyrolysis. They were previously recommended for electrical fires but their use is now prohibited in many countries.

Bromo-chloro-fluoro-methane (BCF) extinguishers

BCF and related substances are liquids which vaporize at about room temperature, forming a dense and heavy vapour which will blanket and extinguish fires of all types with great effectiveness. They are ideal for electrical fires but have limited cooling power. Re-ignition may occur when the vapour disperses. BCF leaves no residue, it is much less toxic than carbon tetrachloride and is generally harmless to equipment and delicate materials.

Buckets of sand

These items are generally provided in chemical laboratories to smother small fires of all types and to contain spilled liquids. Dry sand is particularly effective against minor fires involving alkali metals such as sodium and potassium. Some laboratories also install buckets of sodium carbonate or bicarbonate and these may be used to contain or neutralize spillages of acids.

Asbestos blankets

This type of blanket is intended to smother small fires, particularly those involving burning clothing. Many laboratories are changing over to glass fibre blankets, however, because of the toxic nature of asbestos. If fibreglass is used, it must be sufficiently heavy to fall over the fire and exclude the air supply. Some small blankets that have been widely advertised and sold in the UK are made from such lightweight material that when placed over the fire they may float off in the rising current of hot air and gas.

Automatically operating extinguishers

As mentioned above, automatic water sprinkler systems are not generally recommended for laboratory use but special equipment may be provided for particular areas which are especially hazardous, such as solvent stores and chromatography rooms where large quantities of inflammable solvents are stored or used. To avoid unnecessary damage arising from malfunction of the detector or a false alarm, the laboratory use of this equipment is generally confined to systems which use carbon dioxide or vaporizing liquid as the blanketing agent. Many of these systems incorporate a delay device that gives an audible warning some seconds before the extinguishing agent is discharged, so that staff may escape before they are asphyxiated by the gas or vapour. Once started, the system cannot be stopped, with the result that the entire contents of the cylinder are discharged into the room or store.

FIRE DRILLS

A considerable number of fire authorities will require regular fire drills to be carried out in all buildings which contain laboratories, or in workplaces where more than a very small number of persons work. The purpose of holding regular fire drills is to ensure that all staff are familiar with their particular route of escape, together with the action that is required of them in the event of an emergency. All staff must be aware of the local fire rules and, to this end, fire notices should be posted in prominent places such as halls and corridors, so that staff are constantly reminded of the correct action to be taken in the event of fire. In laboratories, staff are generally required to sound the alarm, close down experiments in progress, shut off or close down fume cupboard ventilation, close all windows or doors, turn off the gas and electricity supplies and make sure that no person is left or trapped in the laboratory or adjoining rooms.

When the alarm is raised, all staff with the exception of those having specific tasks to perform, such as those in fire teams, will be expected to leave the building by the shortest designated escape route and assemble in a specified safe area. In many buildings which have a fairly static population, such as schools, it will be possible for section leaders or departmental heads to check that all their staff have escaped safely by means of a roll call carried out at the assembly point. In other buildings such as universities and research institutes, where the population is likely to fluctuate, it will be necessary to make other arrangements to ensure that all staff have left the building. One method which is widely used is to organize fire teams on each floor, or in each section, whose task when the alarm is given is to search every room individually to ensure that it is clear of all persons and that no one is trapped or is unaware that the alarm has been given. Special arrangements may also be made so that any staff with physical disabilities are assisted from the building.

FIRE ESCAPES

These are provided to ensure that all the occupants of the building, including visitors, have at least one and usually two safe and smoke-free routes of escape in the event of fire. In a new building or where there has been a change of use in an older building, adequate provision of escape routes will have to be made at the design stage in accordance with the local authority's fire and building regulations. Generally, the route of escape will be by means of designated and signposted corridors and stairways constructed of non-combustible materials. To inhibit the spread of fire and smoke, the stairways are protected at every entry by fire-resistant doors of approved design. In single-floor buildings, a number of external doors fitted with 'panic bolt' locks will be required. In an emergency, doors fitted with this type of lock may be easily opened by pressure on an internal horizontal bar, without the need of keys. Windows at ground floor level are not usually regarded as an acceptable means of escape because of the difficulty that less agile persons may encounter in using them.

Local authorities may insist that each laboratory shall have at least two widely spaced doors or one door and a 'kick-out' or other easily removable panel that gives access to another room or corridor. It is important that all doors, removable panels, corridors and stairways on fire escape routes are kept free from any obstruction and that all smoke and fire doors are normally kept closed. Most local authorities require internal fire doors to be fitted with automatic closing devices and that such doors are not held open by means of hooks, bolts or wedges. Doors which close automatically may present considerable difficulty to staff carrying apparatus or pushing trolleys, or where the doors are in constant use. Some authorities will allow the use of electrically operated magnetic catches to hold the doors open. In the event of fire, the sounding of the alarm or the failure of the electrical control circuit, the catches are released and the doors close automatically. In older buildings, external fire escapes may be provided and it is important that these also remain free from obstruction by dustbins, discarded equipment, etc.

The number and size of fire escapes and the width of the designated corridors which give access to them will depend on (a) the size and complexity of the building, (b) the type of work carried out, (c) the number of floors, and (d) the number of persons likely to be present in the building at any given time.

FIRE PREVENTION ADVICE

Local fire brigades and fire prevention associations are generally prepared to give advice and to make recommendations concerning the provision of fire-detecting and fire-fighting equipment. Many of the large suppliers and manufacturers of fire equipment are also willing to help in this way and some of them will inspect and maintain their own and other equipment on a regular contract basis.

Electrical and electronic equipment

Faulty electrical equipment and installations are said to be the largest single cause of fires. Staff may also receive burns and electric shocks from the improper use of electricity. All electrical equipment should be inspected for safety by a suitably qualified person on delivery and subsequently at least once per annum. Many laboratories restrict the issue of plug tops to professionally qualified staff to ensure that the apparatus is properly checked, wired and fused before commissioning, but with the increasing complexity of modern laboratory equipment it is becoming more difficult to find persons with sufficient knowledge and experience to perform this task effectively; but nevertheless it is essential for all equipment to be properly installed and maintained.

In general, electrical apparatus is perfectly safe when properly used, but overloading or overheating may lead to the breakdown of insulation and short circuiting of the electrical supply. Equipment used in damp conditions, such as those prevailing in cold-rooms and glass-washing areas, may start a fire due to flashover; badly maintained high voltage equipment may produce the same result. Sparks from thermostats and switches have been known to cause fires, particularly in instances where solvents are exposed in unventilated spaces such as refrigerators, cold-rooms or other enclosed air systems.

Inflammable solvents must never be centrifuged in open tubes or pots because of the very high risk of fire or explosion resulting from sparks produced by the motor or control equipment. Electric motors with badly worn brushes or commutators are a prolific sources of sparks and, because they are often concealed, may be overlooked when solvent is spilled or exposed. Sources of sparks must also be kept away from charging batteries or other equipment likely to emit oxygen and hydrogen in explosive proportions. Both the authors are aware of an accident in which an experienced technician received severe shock and injury when he attempted to repair a crack in the top of a recently charged battery with Chatterton's compound applied with a hot soldering iron.

High-voltage equipment must be installed and used with extreme care because the consequences of any accident are much more likely to prove fatal. All such equipment, and particularly high-voltage electrophoresis equipment, which may well be water-cooled, must be fitted with interlocking switching so that it is impossible to operate it with the case open and high-voltage points exposed.

Radiation and the use of radioactive substances

The effects of radiation damage are not immediately apparent and when equipment producing ionizing radiation is used or when radioactive substances are handled, there may be no general indication of the danger to which a person may be unwittingly exposed.

The increasing laboratory use of isotopes and radioactive substances in recent years has brought about an increased awareness of the hazards involved. Many laboratories also use X-rays in routine diagnostic and research techniques such as X-ray crystallography. There is danger, not only to the users of such substances and to equipment, but also to colleagues and members of the public who may not be aware of the hazards to which they are being exposed.

In most countries the use of these substances and equipment is well controlled by legislation[4], but the laboratory manager will need to ensure that suitable facilities are provided, that the appropriate rules and regulations are being observed and that staff are adequately trained.

Cylinders of compressed gas

The number of gas cylinders in use in laboratories should be kept to a practicable minimum. They must always be firmly fixed or supported by means of chains or straps and must never be used in situations where the temperature is likely to rise significantly, e.g. near radiators, in direct sunlight or in hot-rooms.

Cylinders of compressed gas must always be used with the appropriate control heads or pressure regulators, together with suitable non-return valves and, where appropriate, flame traps. Control heads, pressure regulators and non-return valves must never be oiled or greased. If connected to any thin-walled metal apparatus (and certainly when connected to any apparatus constructed of glass) there must be provision for automatic pressure release so that the apparatus is not submitted to undue stress.

Cylinders must always be transported in properly constructed cylinder trolleys and must never be dropped. A large cylinder from which the head valve has sheared off will propel itself in the manner of a rocket projectile. The storage of gas cylinders is discussed in Chapter 4.

Centrifuges

Mention has already been made of the danger from sparks and inflammable solvents in connection with the use of centrifuges, but there are several other hazards that laboratory managers need to consider.

All centrifuges, heads, buckets, trunnions, etc., should be regularly inspected to ensure that they have not been damaged or weakened by misuse, wear or corrosion. The buckets and heads, in particular, should be washed, dried and inspected after each use so that corrosion may not occur without the appropriate remedial action being taken.

It is important that centrifuge heads, buckets, bottles and tubes are not subjected to greater stress than they were designed to withstand, and to this end a chart showing the maximum speed of each rotor or head and bucket assembly should be displayed on, or close to, each centrifuge. It should be remembered that the force (g) acting on a tube or rotor increases not only in direct proportion to its distance from the centre of the spindle, but also increases as the square of the spindle speed, and therefore it is particularly hazardous to subject any component to speeds above the recommended maximum.

The smaller and slower centrifuges and heads are generally designed and built to have a fairly large reserve of strength so that, in normal use, they are able to withstand a certain amount of abuse. With the large-capacity and ultra-high-speed machines and particularly so with the modern high-speed centrifuge, the material from which the heads or rotors are constructed is subjected to stresses and strains which bring it perilously near to its ultimate tensile strength. For this reason, any scratches, scores or signs of the slightest corrosion that may appear in an ultra-high-speed head or rotor must be dealt with immediately; this is usually done by returning the head to the manufacturer or the manufacturer's agent for inspection and possible repair. Only limited repairs are possible to ultra-high-speed heads, these usually being confined to the fitting of new centre pieces or cleaning and re-anodizing. All too frequently, an expensive rotor is ruined and has to be scrapped because insufficient attention has been paid to its regular cleaning and drying after use.

Most ultra-high-speed heads have a limited life. Some manufacturers recommend that the head be derated after, say, a period of four years, irrespective of how much it has been used, whilst others recommend that the head be derated or scrapped after a given number of accelerations; still others recommend that the head be derated after a defined total number of revolutions.

The high-speed centrifuge not only costs a great deal to purchase and a great deal to maintain properly, but there may also be very heavy expenses involved in the almost unavoidable replacement of heads and rotors. It is absolutely necessary for the user to ensure that, where necessary, centrifuge tubes and buckets are accurately balanced. Failure to carry out this procedure may give rise to increased wear on bearings and spindles and, in severe cases, may cause the centrifuge to overturn.

On smaller machines not fitted with automatic braking or interlocking lids, staff must be discouraged from increasing the braking effort by applying their hand to the head retaining nut whilst the head is rotating. Occasionally whilst the centrifuge is running, a glass tube or bottle will break and small pieces of glass will escape into the centrifuge bowl. Should this occur the centrifuge must be stopped, disconnected from the power supply and the bowl thoroughly cleaned out. Failure to do so may lead to increased wear on the top bearing and abrasion of the centrifuge head.

Cryogenic substances

Many laboratories use liquid gases or cryogenic mixtures, such as solid carbon dioxide and ethyl alcohol, in order to produce low temperatures in cooling baths or traps. Liquid nitrogen is also widely used in the storage and preservation of biological materials. It is important to remember that these liquids and mixtures may produce very painful and severe burns and destruction of tissue if allowed to come into contact with the body, and therefore suitable protective clothing must always be worn when handling them. Gloves, face masks and aprons are usually provided. Rubber boots, if provided, should be worn inside trousers to prevent spilled liquid falling into the boot which may be extremely difficult to remove quickly in the event of a spillage. For the same reasons it is inadvisable to wear gauntlets.

There are a number of precautions that need to be observed when using cryogenic substances[9].

(1) Liquid nitrogen, solid carbon dioxide (or mixtures containing it) should not be used or stored in confined spaces or rooms with little or no ventilation, since there is considerable risk that the air in such a room will be rapidly depleted of oxygen by the release of large volumes of relatively inert gas. It is not unknown for laboratory workers to store blocks of solid carbon dioxide in cold-rooms, which generally have very little ventilation. This practice will very soon produce a suffocating atmosphere in the room.

(2) Liquefied gases and other very cold liquids must not be poured into unsuitable containers. The thermal shock produced may well be sufficient to shatter the container; many of the vacuum flasks produced for domestic purposes, for example, are incapable of withstanding the very sudden lowering of their temperature when liquid nitrogen is poured into them.

(3) Liquefied gases must not be stored in containers or vessels that are not freely vented to the atmosphere. There is a danger that gas released by evaporation from the liquid will be unable to escape from the vessel at a sufficient rate to prevent the vessel becoming pressurized and subsequently exploding. This is particularly important with narrow-necked, metal, vacuum-insulated storage containers. In humid conditions, it is possible that ice may form at the mouth of the container and this would severely restrict the outflow of escaping gas.

(4) Liquid nitrogen is generally much more expensive than the liquid air from which it is produced commercially by a process of fractional distillation. The use of liquid air should be avoided whenever possible, because there is the risk that on prolonged storage the nitrogen will boil off preferentially, leaving a mixture with an increased oxygen content. Substances which burn only with difficulty in air will burn with vigour in an oxygen-enriched atmosphere and sometimes explosively when soaked in liquid oxygen.

Physical injuries

The incidence of physical injury to laboratory staff may be reduced by good initial design, by the provision of well-lighted rooms, corridors and staircases, etc., and by the use of non-slip floor surfaces. Many materials commonly used for laboratory flooring, including some grades of PVC, may be perfectly safe when dry but may be extremely slippery when wet. The level of injury may be still further reduced by good housekeeping which will ensure that unused pieces of equipment, rubbish, etc., are not allowed to obstruct gangways and corridors.

Discipline also has an important part to play in the reduction of accidents, in that staff should be discouraged from bustling and running in laboratories and corridors. They must also be prevented so far as is possible, from indulging in the wilder forms of practical jokes and clownish behaviour.

Staff must be trained in the correct use of apparatus and equipment, especially workshop machinery. Junior staff must not be allowed to work unsupervised in workshops or laboratories (see Chapter 6). They must also be taught the correct methods to be used when lifting or carrying heavy objects. In the UK, the Health and Safety at Work Act[1] may shortly place a limit on the weight which an adult person of 18 years or more is required to lift or carry.

Suitable equipment should be provided for the safe transport of hazardous substances or heavy equipment. Staff must be discouraged from using laboratory stools and chairs and other makeshift aids to reach high shelves.

High-vacuum equipment should be properly maintained and regularly examined and tested in accordance with any legal requirements. It is also essential to comply with any conditions imposed by an insurance company, e.g. most insurance companies require that any autoclaves and pressure vessels for which they issue insurance cover should be thoroughly examined and tested at least once per annum.

Broken glass, scrap metal, used hypodermic needles and other sharp items must be segregated from general laboratory waste. Staff must also be forbidden to place broken glass, etc., into waste paper containers. It is the duty of the laboratory manager to ensure that all waste generated in the laboratory is disposed of safely.

First-aid equipment and materials should be provided in accordance with any legal obligations, such as those required by the Factories Acts[3] and the Offices, Shops and Railway Premises Act[6]. The scale on which such equipment and materials are provided is generally a reflection of the number of staff in the establishment and the hazards likely to be encountered. It should be remembered that first aid is intended to deal expeditiously with minor physical injuries and, in the case of more serious accidents, to preserve the victim from further harm until professional help and treatment can be obtained.

Chemicals

There are many hazards associated with the use and handling of chemicals. Much has been written in an effort to reduce or eliminate the effect of them on staff, and to reduce the cost and inconvenience of damage to equipment and property. It is reasonable to expect, although it may not always be found so in practice, that chemists and other specialists will have received in their training, detailed instructions in the safe use of the substances that they use and handle in their daily work. For further information of this type, the reader is referred to the lists of References and Further Reading at the end of the chapter.

Chemicals are used by workers in many fields, some of whom may have had little or no training in chemistry. Many substances in common use are highly toxic and their use by young or inexperienced persons (particularly in schools) needs to be carefully supervised and controlled. Whilst having some knowledge of the detail, however, the laboratory manager needs to be more concerned with the broader organizational aspects of the problem.

Occupational hygiene

In all laboratories in which toxic or infectious substances are used, a high standard of cleanliness and personal hygiene must be maintained.

Eating, drinking, smoking, the application of cosmetics, licking of labels, mouth pipetting, chewing of pencils, biting of fingernails, or any activity which could lead to the ingestion of toxic substances, should be prohibited. No doubt the enforcement of such a rule will be difficult and unpopular but this does not lessen its necessity. The accomplishment of a task of this nature will be found to be less traumatic if the senior staff can be persuaded to set a good example. Since chemicals may also enter the body by inhalation and by absorption through the skin and eyes, particular care must be exercised when weighing or transferring toxic substances in dry powder form from one container to another.

Laboratories in which toxic substances are handled should be well ventilated and provided with sufficient properly designed and constructed fume cupboards, suitable washing facilities with hot and cold water, and an adequate supply of disposable paper towels.

For staff working with highly corrosive chemicals, it will be necessary to provide goggles or face masks and rubber or plastic gloves. If the work being carried out involves the use of more than small amounts of corrosive substances, it may be necessary to provide rubber or plastic aprons and boots. Emergency water drench showers are also very useful for dealing with a major spillage of corrosive substances on to the clothes or person. All staff should be trained so that full use is made of the facilities provided for any such emergencies. Supervisors must ensure that all spillages are effectively cleared up immediately they occur and that a suitable standard of cleanliness is maintained at all times. It is important that laboratory coats, overalls and gowns are worn by all laboratory staff whilst working and that they are removed before entering restaurants and other rooms where food is consumed. Examination of an old laboratory coat should convince most laboratory workers of the necessity to protect their personal clothing from at least some of the corrosive material that is encountered in laboratory work.

Dermatitis and skin reactions

Many substances, such as chromium salts[10], chlorinated hydrocarbons, phenolic compounds and animal hair and dust, will produce skin reactions and lesions which are not only of unpleasant appearance but are often extremely painful and slow to heal. Sensitivity to these substances varies enormously between individuals. Many persons are liable to become so sensitized that after receiving a few fairly short initial exposures to the substance their response to a given dose or exposure becomes much greater and the result more painful. If they are isolated from the substance they will generally recover, but a subsequent brief exposure may well produce an immediate and severe response. At this stage, they may have to be removed permanently from the work that brings them into contact with the substance to which they have become sensitized.

Many laboratory staff develop allergies to animal hair and therefore find it difficult to work with certain species of animals. Some become so sensitized that within minutes of entering a room containing an animal of the particular species they suffer from acute symptoms and become quite unfit to continue with their work. Other staff are, or may become, allergic to the dust produced by locusts and other insects.

Chromic acid which was formerly widely used for cleaning laboratory glassware is still, regrettably, a common cause of contact dermatitis. Prolonged exposure of the hands and forearms to chromates will produce deep, sharply defined ulcers that are slow to heal[10]. In many laboratories, chromic acid has now been replaced by modern powerful detergents specially formulated for cleaning glassware and their use has done much towards reducing the incidence of contact dermatitis. Because of the nature of their work, laboratory glassware cleaners need to be especially protected from the effects of these harmful substances. If they are allowed to put their bare hands into hot water containing detergent, their skin may become defatted and thus be more easily penetrated by the causative agents. The laboratory manager must insist that the glassware cleaners wear rubber or plastic gloves when they are working and that they report any sign of skin irritation immediately.

Toxic substances and threshold limit values

Some countries, notably the UK and the USA, have introduced legislation or regulations that define the maximum concentration of a substance in air to which a person may be exposed whilst at work[11]. Threshold limit values (TLVs) refer to a time-weighted concentration for a 7- or 8-hour working day or a 40-hour working week. TLVs are expressed as parts per million (ppm) or milligrams per cubic metre (mg/m^3). Time-weighted averages generally permit excursions for a limited period above the limit, provided that it is compensated for by excursions below the limit. Some substances have ceiling limits above which no excursion is permitted. The tables for TLVs are based on the best available information from industrial experience and from experimental human and animal studies.

Study of the TLV tables will show surprisingly low limits for many substances which are commonly found in laboratory atmospheres, e.g. chlorine 0.1, formaldehyde 2.0, hydrogen sulphide 10, pyridine 5 (all in ppm). A number of substances will be found to have TLVs at about, or below, the concentration at which many persons are able to detect them by their sense of smell, e.g. carbon tetrachloride has a TLV of 10 ppm, but most people can only detect levels above 50 ppm. Hydrogen sulphide also has a TLV of 10 ppm, and although many persons are able to detect as little as 0.13 ppm, they are, on exposure, likely to become· rapidly fatigued so that their response is then above the TLV. It follows that a sense of smell is a useful, but not a very reliable, guide to the concentration of the substance present. It should be noted that it is possible for two substances of modest toxicity to combine chemically or react to produce an extremely toxic product, e.g. the TLV of chlorine is 0.1 and that of formaldehyde is 2.0. It is possible, but not very likely, that under certain conditions these two substances could react together to form the extremely carcinogenic substance bis-chloro-methyl-ether, which has a TLV of 0.001 ppm.

Carcinogens

A number of substances have been associated with an increased risk of the development of neoplastic disease in both man and animals. More are suspected of carcinogenic action and many more remain as yet untested for this type of activity. A number of substances whose use has been suggested as a safe substitute for known or suspected carcinogens have later been found to be almost as active in this respect as the original material.

With new substances being added almost daily to the list of compounds known to have, or suspected of having, carcinogenic properties, it is imperative that all laboratory staff should be aware of the hazards and that adequate precautions are taken whenever these substances are used or handled in the laboratory.

It has been suggested by Howe[12] that, as far as animal experiments are concerned, it is possible to grade the available evidence about carcinogens into three arbitrary categories:

(1) Potent carcinogens — strong, proven carcinogens associated with high incidence of cancer.
(2) Carcinogenic — proven carcinogens with moderate or weak activity.
(3) Suspect carcinogens — inconclusive evidence of carcinogenicity or untested compounds structurally related to known carcinogens.

Compounds falling within the first group require precautions and facilities that are unlikely to be available or possible in many laboratories. The manufacture, importation and sometimes the use of these substances has been restricted or prohibited in many countries. In other cases, the legal requirements imposed to ensure their safe manufacture or use have been so severe as to render the process or use uneconomic. Work with substances falling into the second and third groups will generally require facilities and precautions similar to those provided and taken for work involving the use of radioactive isotopes.

In general, the risk of developing neoplastic disease from working with carcinogenic substances is in proportion to (a) the length and frequency of exposure, and (b) the amount or concentration of the substance to which one is exposed.

Irrespective of the type and quantity of equipment that is provided for the safe handling of carcinogenic substances, and of the rigour with which the precautionary measure are enforced, it is essential that a very high standard of occupational hygiene is maintained when and wherever they are used or handled.

Many codes of practice have been published for the safe use of carcinogens and a number are referred to at the end of this chapter[13,14]. Without doubt, the best advice is to avoid the use of these substances altogether and, if this is not possible or reasonably practicable, to substitute safer and less hazardous compounds whenever possible. Because the development of cancer is often a prolonged and insidious process and because at present the carcinomatous process once commenced is irreversible, young persons should not be required to use carcinogenic substances.

Among the substances which are known to be carcinogenic are asbestos, alpha- and beta-naphthylamine, benzidine, the compounds of diazomethane, the nitrosamines, the nitrophenols and many dyestuffs and stains.

There are a few substances which are thought to be co-carcinogens, in that when used by themselves they are safe, but there is evidence to suppose that they have the ability to enhance the tumour-producing effect of some carcinogenic substances. There is also the possibility of synergistic action — in a few cases two carcinogens, when used together in amounts that would by themselves be relatively innocuous, are known to give a high incidence of tumours in experimental animals.

After use, carcinogens and materials such as filter papers which have been contaminated with them are best destroyed by chemical action. If this is not possible, then they may be disposed of by incineration under carefully controlled conditions, although this may be a difficult operation, particularly where the substance is volatile and may escape up the chimney before being completely destroyed. Many local authorities refuse to allow the dumping of waste contaminated with carcinogens at their tips and some impose special conditions before allowing such waste to be dumped in the area for which they are responsible.

It is the duty of the laboratory manager to ensure that all dangerous substances arising from work in his laboratories are disposed of safely, and with due regard to any local or other legal requirement.

Bacteria, viruses and other biohazards

In recent years there has been an increase in the number of laboratories in which microbiological materials are handled. There has also been a parallel increase in the number of staff engaged in this type of laboratory activity. Many of these workers have entered this particular field after having received their primary training in other (but not necessarily very closely related) disciplines. Work with micro-organisms can be extremely satisfying and the very recent work in the field of genetic manipulation has attracted a number of persons into this branch of science. Some of these may have previously received very little training in dealing with the special hazards that are likely to be encountered in microbiology. It matters not that a particular worker is only interested in the biochemical or biophysical, as distinct from the purely biological, aspects of a problem. If biologically hazardous material is being used or handled, then the appropriate safety measures must be prescribed and, if necessary, enforced.

Many individual institutes and research organizations have developed or suggested, and a few have published, codes of practice for the safe handling of infective or potentially biologically hazardous material. A number of reports by expert committees, many of them sponsored by government or semi-government agencies, have given guidance on the safety precautions to be observed by laboratory workers[15-17].

For a fairly short period there was a voluntary embargo entered into by many workers and organizations interested in genetic manipulation, so that the broader aspects and implications of the work could be considered before laboratory staff and the population at large were exposed to ill-defined and largely unknown hazards. In the UK, this type of work has been resumed and is now monitored by the Genetic Manipulation Advisory Group (GMAG). It may be expected that for work with the most hazardous biologically active materials, legislation and some form of government control will soon follow.

Many micro-organisms are pathogenic to man, and others may be able to cause widespread damage to the environment by bringing about the destruction of livestock and crops if they are allowed to escape and are able to proliferate. It is important, therefore, that every laboratory in which biologically hazardous or potentially biologically hazardous material is used or handled, should produce or adopt an appropriate code of practice for the work that is to be undertaken. There must be adequate supervision to ensure that the spirit as well as the letter of the code is followed by all staff.

In a draft report dealing with the prevention of laboratory acquired infection, Howie[16] has suggested that pathogenic micro-organisms, infectious agents, etc., may be classified into three major groups according to the hazards they present and to the minimal safety conditions for handling them. These groups are:

(1) Category A pathogens, which include organisms, viruses and materials such as Simian Herpes (B) virus, Lassa fever, Marburg and smallpox viruses, etc., which are extremely hazardous to laboratory workers and which may cause serious epidemic disease. These will require the most stringent conditions for their containment.

(2) Category B pathogens, which include organisms, viruses and materials such as tubercle bacilli, brain tissue and brain extracts containing the causative agent of Creutzfeldt–Jacob's syndrome, and work involving more than the occasional routine isolation of *Clostridium botulinum, Mycobacterium tuberculosis, Salmonella typhi,* pathogenic amoebae, etc. It is considered that work with

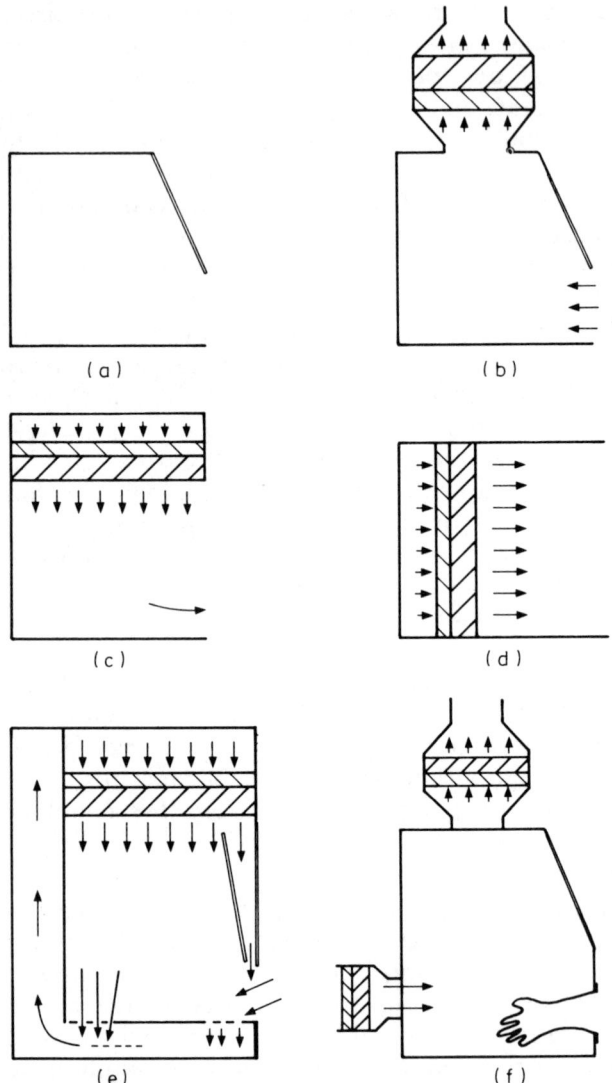

Figure 7.1. Clean-air and microbiological safety cabinets: (a) microbiological workbox, which gives a measure of protection to the work; (b) exhaust protective cabinet, which protects the operator but not the work; (c) clean-air cabinet with vertical laminar flow; (d) clean-air cabinet with horizontal laminar flow; (e) recirculating biological safety cabinet, which protects both the operator and the work; (f) microbiological glove box — this is the only type of cabinet that may be used for handling highly infectious or extremely hazardous materials

material in this category will offer special hazards to laboratory workers and that special accommodation or conditions for containment must be provided.

(3) Category C pathogens, which include organisms, viruses and materials not listed in categories A and B that offer no special potential hazard to laboratory workers but which may be contained by the use of good technique.

The report of the Howie Working Party[16] is primarily concerned with organisms, etc., that are pathogenic to man and which may be found in clinical laboratories. Other organisms and materials that may be hazardous to animals or to man's environment and only indirectly to man, may need similar consideration. They should first be classified, so that their handling and containment may be related to hazards presented by their use.

A number of references are provided at the end of this chapter which may assist in the compilation of a suitable code of practice for many types of biological work. As with other safety measures that may be recommended, it is important to remember that the provision of elaborate codes of practice and of expensive equipment and facilities such as biological safety cabinets and glove boxes, etc., should not be allowed to become an excuse for sloppy workmanship (*Figure 7.1*). There is no substitute for adequate training in basic laboratory techniques and there should be a constant awareness of the hazards, however insignificant, that are likely to be encountered in one's daily tasks. Laboratory managers have a special responsibility to ensure that all of their staff, both scientific and technical, for whom they are responsible are proficient in the basic skills of their craft[18].

REFERENCES

1. The Health and Safety at Work etc. Act 1974. HMSO, London
2. *Safety and Health at Work Report on the Committee 1970–72* (Chairman Lord Robens). HMSO, London.
3. The Factories Acts 1937 and 1961. HMSO, London.
4. The Radioactive Substances Act 1960. HMSO, London
5. The Petroleum (Consolidation) Act 1928. HMSO, London
6. The Offices, Shops and Railway Premises Act 1974. HMSO, London
7. Cooke, A. J. D., *A Guide to Laboratory Law.* Butterworths, London (1976)
8. The Fire Precautions Act 1971. HMSO, London

9. *Safety Precautions for the Handling of Cryogenic Liquids.* Edwards High Vacuum Ltd, Crawley, Sussex (1978)
10. *Chromium Health & Safety Precautions,* Guidance Note EH/2. HMSO, London (1976)
11. *Threshold Limit Values for 1976,* Guidance Note EH 15/76. HMSO, London (1976)
12. Howe, J.R., 'A method of recognising carcinogens in the laboratory', *Laboratory Practice,* 24, 457—567 (1975)
13. *Precautions for Laboratory Workers who Handle Carcinogenic Aromatic Amines.* The Chester Beatty Research Institute, London (1966)
14. *Code of Practice for the Handling of Carcinogens.* Medical Research Council, London (in preparation)
15. Collins, C.H., Hartley, E.G. and Pilsworth, R., *The Prevention of Laboratory Acquired Infection,* PHLS Monograph No. 6. HMSO, London (1974)
16. *Report of the Working Party to Formulate a Code of Practice for the Prevention of Infection in Clinical Laboratories* (Chairman Sir James Howie). Department of Health & Social Security; HMSO, London (1978)
17. *Report of the Working Party on the Practice of Genetic Manipulation* (Chairman Sir Robert Williams). HMSO, London (1976)
18. Grover, F. and Wallace, P., *Nature,* 272, 391—392 (1978)

FURTHER READING

A Guide to Recognising and Handling Carcinogens in a Veterinary Laboratory. Agricultural Development and Advisory Service, Central Veterinary Laboratory, Weybridge, Surrey (1975)

Guy, K., *Laboratory Organization and Administration.* Butterworths, London (1973)

Hartree, E. and Booth, V. (Eds.), *Safety in Biological Laboratories.* The Biochemical Society, London (1977)

Hellman, A., Oxmen, M.N. and Pollack, B., *Biohazards in Biological Research.* Cold Spring Harbor Laboratory, New York (1973)

Muir, G.D., *Hazards in the Chemical Laboratory,* 2nd edn. The Chemical Society, London (1977)

Sax, N.I., *Dangerous Properties of Industrial Materials,* 4th edn. Van Nostrand, New York (1975)

8
Maintenance of Laboratory Premises and Equipment

Like any other buildings, laboratories deteriorate with time and use. They are subjected to continuous wear and attack by corrosive materials, carry a great deal of traffic and are, perhaps, not always treated with sufficient respect by the staff. Equipment, too, is often used in conditions which are far from ideal, is expected to run continuously without faltering and is treated with very little consideration. Faults develop which not only put apparatus out of action but can also make it unsafe to use, e.g. an electric oven with a defective thermostat or an animal cage with a broken latch.

Regular inspection and maintenance will not completely prevent breakdown but it will certainly reduce the chances of such problems occurring. The need for regular and efficient maintenance cannot be overstressed, both in terms of safety and economy. Furthermore, a shabby, dirty laboratory, having apparatus which is inaccurate and unreliable, could lead to a general despondency among the staff and eventually to an acceptance of lowered standards. A situation such as this becomes difficult, if not impossible, to cure.

Planned maintenance

This system is designed to give regular inspection and maintenance of the premises, fittings and major items of equipment. It is carried out at intervals, say once or twice a year, either by the staff of the organization or by outside contractors. Facilities such as the lifts, cold storage rooms, safety equipment, heating and ventilating plant, etc., are maintained by specialist firms, preferably those who supplied or installed the equipment.

Major apparatus is generally made the subject of service contracts entered into with the various companies which either manufactured or supplied the apparatus. This approach has a number of advantages over the 'do-it-yourself' alternative. The manufacturer should employ service engineers who are familiar with the equipment and with the faults which may occur and should carry stocks of parts which may need replacement. They should also have the necessary equipment for testing, setting up and calibrating — an expensive and specialized process which would be uneconomical for the laboratory to attempt. Furthermore, the cost of an annual contract may well include spare parts and emergency calls, apart from the regular visits agreed upon. Although the cost of such a contract may be fairly high, it can be included in the annual budget with the knowledge that no further expenditure will be necessary during the period of the agreement, normally 12 months.

Where similar apparatus of several different makes exist in the laboratories, it is sometimes possible for one company to service them all (balances and microscopes are good examples). By having all the work carried out by one engineer in one visit, costs may be substantially reduced.

Visiting service engineers must be made aware of any particular hazard which may be present, both in the apparatus they are to service and in the room (or rooms) where they are working[1]. Every possible effort must be made to remove such hazards, e.g. a centrifuge used with bio-hazardous materials would be decontaminated by a responsible person before being handed over for servicing. In some laboratories, service engineers are not permitted to work without a member of the staff being in attendance.

It is not always realized that calling in a service engineer can be very costly. Charges are based on time spent, including travelling time, plus spare parts used. One occasionally encounters an engineer who is not as competent as one would hope and who may spend an unreasonable amount of time locating and rectifying faults. Justifiable complaints relating to such matters should be taken up with the company concerned, which may be prepared to adjust their charges accordingly.

Routine servicing does not often reveal major defects but even minor faults can at best reduce the efficiency of the equipment if left uncorrected and may in time develop into serious faults which could result in the apparatus becoming so badly damaged that it would be beyond repair or make it the cause of a serious accident. For example, minor corrosion of a high-speed centrifuge head can be rectified, but if left untreated will result in further corrosion making costly replacement necessary or leading perhaps to failure of the head when it is running at maximum speed (see Chapter 7). An analytical instrument marginally out of adjustment may give false results without the user being aware of any error.

Some items of equipment should be regularly maintained by the staff. Work such as descaling stills, replacing batteries and lamps in apparatus, routine adjustment of equipment, etc., does not warrant calling in service engineers and several members of the staff can have these tasks delegated to them. This work should, of course, be limited to fairly simple and straightforward apparatus and the manager must be confident that the work is within the capabilities of the staff concerned. Delicate or complex apparatus can easily be damaged beyond repair by the untrained but it is not difficult to establish whether or not a particular employee has the ability to undertake such work.

If the organization has the benefit of a workshop with skilled staff, a considerable amount of repair and maintenance work would be done by them (see Chapter 6).

INSPECTION OF PREMISES AND EQUIPMENT

In the course of his normal duties, the laboratory manager will of necessity circulate throughout the building and in so doing will observe obvious defects, such as damaged floor coverings and cracked windows, and he will make a note of these and take the necessary remedial action. Other minor defects will be reported by the occupants of the laboratory.

There should, however, be a regular system of inspection at such intervals as may be required to ensure that nothing is overlooked. An inspection should be carried out prior to the preparation of the annual budget, so that costs of repairs and upkeep may be included in the financial estimates[2]. Each room should be visited, particularly those not in general use, and notes made regarding the condition of decoration, lighting units, fume cupboards, bench services, etc.

The importance of good lighting in the laboratory was discussed in Chapter 1. During their useful life, fluorescent lamps depreciate in light output by about 20%. Dust and dirt accumulating on reflectors and diffusers can further reduce the light output to as little as 50% of the original specification, and some units in industrial areas may need cleaning at intervals of about 4—6 months to maintain their efficiency[3]. Where the laboratory is part of some larger organization, defects are reported to the department concerned and arrangements are made for their correction.

Defects which could lead to further damage, such as leaking sink wastes, or faults which are hazardous, such as damaged electrical outlets, must be treated as matters of urgency. Other work can be done at some mutually convenient time or may be deferred until there is less demand on the laboratories and maintenance staff. In a teaching department, for example, work which is not urgent could be done during vacations. Laboratories which have no maintenance staff at their disposal must usually rely on an outside contractor whose work is known to be of an acceptable standard and who can carry out urgent work with the minimum of delay.

Inspection of apparatus should be done in much the same way. Although major faults are normally reported by the user, minor defects are often ignored. Correction of these minor faults results in greater accuracy and improved reliability and safety. It should be remembered that apparatus only fails when it is in use and that is precisely when it is most needed. Inspection is also a useful guide to the care (or otherwise) with which apparatus is used and it provides an opportunity of considering whether or not some items should be replaced rather than repaired. Even the best available equipment may fail unpredictably but failure due to neglect cannot be accepted.

Replacement of apparatus

Restocking laboratories with consumable items such as glassware should be a continuous process, but replacement of major equipment is usually effected when required or, alternatively, at annual intervals when funds may be available.

The useful life of apparatus is a function of several factors, i.e. the type of equipment, its quality, the amount of use it has and the conditions under which it is used. A carefully used balance may, for example, last for 20 years or more but an electronic instrument used continuously in a corrosive atmosphere may be written off in one year. The laboratory manager must often decide between the merits of repairing or replacing a major piece of apparatus. Clearly, a time must come when repairs cannot be expected to keep the equipment in good order for very long, and when repair follows repair at progressively shorter and shorter intervals, costs can mount to a figure in excess of replacement cost. Equipment is rapidly superseded by improved models and it is generally more satisfactory to replace an old item with an updated version if the cost difference between repair and replacement is not excessive.

With today's high cost of labour it is uneconomical to employ staff to repair or service small inexpensive items. For example, a technician who spends two hours repairing a simple time switch probably costs his department more than the purchase price of a new unit. Obviously, if a replacement unit was not obtainable or was needed immediately, a repair would be the only possible solution and the time would have been well spent.

Cleaning of premises

This work may be done by employees of the organization or by contractors. There are certain advantages in employing one's own staff, since supervision becomes easier and staff turnover tends to be lower than with contractors. The main advantage in employing contractors, however, is that they provide all the staff and equipment and thus relieve the laboratory manager of this burden. Staff are also covered during holidays or time off due to sickness, etc., and supervision is carried out by the management of the company.

Cleaning staff should be supplied with a detailed schedule of work, which is divided into tasks to be done daily, weekly, monthly and at longer intervals, usually quarterly or annually. Daily tasks should include the cleaning of toilets, corridor and entrance hall floors, office desks and floors, laboratory floors and lifts and staircases. It should be noted that the cleaning of benches is not usually included as it is more satisfactory, and certainly safer, for the laboratory staff to carry out this work themselves.

Weekly tasks may include washing down doors and window frames, window cleaning, washing and polishing floors, polishing metal fittings such as door handles and paying special attention to items such as staircases or rooms with limited access. Other work of a less urgent nature, such as cleaning light fittings, should be done at less frequent intervals. Extra work may be required at irregular intervals, such as that which arises after redecorating, and this may be charged separately.

Cleaners should not, of course, be allowed into rooms where major hazards exist, such as radioactive laboratories or rooms where pathogenic organisms are known to be present, and the cleaning of these areas should be undertaken by the laboratory staff who are familiar with the hazards. It is essential to establish with the company that their staff are adequately insured against any risk likely to be encountered in the building. The presence of cleaners interferes unavoidably with the laboratory's work and they are therefore usually employed out of normal working hours. If they are involved in an accident, there may be no member of the laboratory staff on duty who can take the appropriate action. One should, therefore, ensure that a responsible person can be contacted in an emergency.

It is an unfortunate fact that cleaners are often suspected of theft — a suspicion which is without any factual foundation. Knowing that their position is a vulnerable one, they are at least as honest as most, but are frequently made scapegoats for others. If an occasion arises where a cleaner is suspected, the matter should be taken up with the company and with the security staff and not with the employee concerned.

Decoration

This should not be allowed to deteriorate to a point where it becomes an expensive project. Frequently, washing down and cleaning is all that will be required but redecorating is usually necessary every 3—5 years depending on the purpose of the room. Generally, laboratories and circulation areas require redecorating every three years, whilst for offices, stores, etc., a five-year interval is usually sufficient. It is a good plan to arrange for this work to be carried out over a three-year period with approximately one-third of the work being carried out each year. In this way, the interference with the work is reduced to a minimum. Although some decorators are prepared to work with large items moved to the centre of the rooms and covered with protective sheets, most will require rooms to be emptied completely. In these circumstances finding somewhere to store the contents of the rooms can be a problem. If corridors are used for this purpose it is essential not to obstruct areas which may be needed as emergency escape routes. Redecorating the entire building in one session is clearly an impossible task.

If the work is to be carried out by contractors it will be necessary to prepare a detailed specification to avoid misunderstanding. The specification should include not only details of the preparation to be done, such as washing down, filling cracks and stripping old paint, but also the type of paint and number of coats to be applied. It is not unknown for decorators to apply one coat of poor-quality paint to a room without any form of preparation, thus resulting in a totally unsatisfactory job. A work schedule should also be supplied which will enable the manager to organize room clearing and rehousing of equipment and staff to enable essential work to continue. As the time taken to decorate a given room is to some extent dependent on the weather and other factors, a few days should be left in reserve to offset against unplanned delays. Decorators may require the use of a locked store for their materials and tools and this point should be established before work is started. Most paints are flammable and must be stored accordingly.

When moving back into the laboratory, the opportunity should be taken of disposing of any unwanted or obsolete materials and apparatus. Thorough cleaning of equipment, etc., should be carried out before reoccupation.

The selection of colours and finishes is dealt with in Chapter 1.

Building repairs

If the laboratory is part of an organization employing maintenance staff, much repair work will be undertaken by the appropriate department, but failing this a building contractor will be employed. In neither case should repair work (nor even inspections) be permitted in rooms where a hazard is known to exist. It is essential that laboratories in which pathogenic organisms, radioactive substances or highly toxic materials are present must be decontaminated beforehand by competent staff. If repair work is likely to be costly, a discussion with all concerned may reveal a less expensive and equally satisfactory alternative approach.

Urgent work must, of course, be carried out as soon as possible but other work can usually be postponed until a convenient date is agreed. It may be necessary to close a room down, in which case the estimated date for completion of the work should be established. If work cannot be interrupted for more than a brief period, the staff and equipment will have to be temporarily rehoused. In the event of the breakdown of a cold-store or hot-room, materials will need to be accommodated elsewhere with the minimum of delay. Services may need to be shut down and this may also affect supplies to other rooms in the building. It is sometimes possible for temporary services to be connected where it is essential to maintain supplies.

Major repair work or alterations must cause some upheaval in the normal running of the laboratories and therefore require considerable forward planning. It may be necessary to employ an architect and to call in several firms to quote for the work or for certain sections of the work. Detailed plans and specifications must be prepared and discussed with all concerned. Essential work may need to be rehoused, access will be required for men and materials and the noise and dirt produced may make it impossible for some types of work to continue. The cutting out of old timber, for example, invariably releases mould spores into the atmosphere, which will contaminate cultures in a nearby laboratory engaged in aseptic work. If possible, the building work should be scheduled for a period when the laboratory is least busy. It is sometimes convenient to arrange for the work to be done in several phases so that one part of the building can be reoccupied before the next phase is started. It is in dealing with problems such as these that critical path analysis can be of considerable help to laboratory manager and building contractor alike (see Chapter 10).

REFERENCES

1. The Health and Safety at Work etc. Act 1974. HMSO, London
2. Guy, K., *Laboratory Organisation and Administration,* 2nd edn. Butterworths, London (1973)
3. *Lighting in Offices, Shops and Railway Premises,* Booklet No. 39, Health and Safety Executive; HMSO, London (1976)

9
Automation in the Laboratory

The word 'automation', in the laboratory context, usually refers to the application of an analytical technique in such a way that it can be carried out by mechanical means with the minimum of human intervention. It is difficult to establish when automation was first applied to laboratory work but, in the past 20 years or so, an increasing number of mechanical aids have been employed. Equipment of a relatively simple nature, such as fraction collectors and diluters, have been in use for a considerable time and from such elementary mechanisms have evolved the complex and versatile instruments of today.

There is an ever-increasing demand for laboratory-produced data, particularly in research work and clinical investigation[1]. At one time researchers were content with a few experimental results but today they require tests to be carried out on large numbers of samples, the results being subjected to statistical analysis. Pathology laboratories are called upon to handle more and more samples from an ever-increasing number of patients as clinicians are becoming more aware of the value of laboratory tests as an aid to diagnosis and treatment. (In the UK, technicians employed by the hospital pathology laboratories are each carrying out about 4000 tests annually.)

Modern equipment is capable of measuring samples, carrying out numerous different tests, analysing the results, presenting the data in printed form and making recommendations on any further action necessary. It may have built-in self-testing facilities and the ability to diagnose its own faults in the event of incorrect operation. It may be coupled to a data storage system from which results can be retrieved at a later stage or kept as a permanent record.

The laboratory manager may be called upon to decide whether or not some degree of automation would be an asset to his organization and he should, therefore, be conversant with the factors to be considered in making such a decision. There is no doubt that automation increases the work output in relation to the number of staff employed. The equipment is capable of operating continuously, and is stopped only for maintenance or reloading with samples and reagents. If hazardous materials are involved, such as radioisotopes or highly toxic reagents, the operator is not exposed to these as would be the case when using manual methods. Staff become available to do other work and the boredom which may result from doing identical tests over and over again is relieved. Inaccuracies and mistakes due to fatigue are eliminated and the need for occasional late working to finish a batch of tests is removed. Depending on the design of the equipment, it may not be necessary to carry out any sample preparation or pretreatment or even to load the apparatus with accurately measured samples. Some designs allow simple reprogramming from one test system to another with the minimum of delay. Most analytical procedures produce inaccuracies in their results and these will vary from one worker to another. Although automated techniques also have in-built errors, they remain reasonably constant for a given item of equipment.

Automation, however, may have certain disadvantages both in terms of standard of work and detrimental effect on staff. The equipment is usually extremely expensive, may require special accommodation and will probably need operators who are trained to use it. A well-tried technique may be unusable and new methods may have to be introduced. Limits of sensitivity and accuracy may be less satisfactory than with traditional methods and the necessary reagents may be more expensive. The equipment will, of necessity, be somewhat complex in its design and construction and servicing charges are therefore likely to be high. Special containers, such as disposable sample cups or vials, may be required and the cost of these can raise the total cost of each test to a higher level than expected. The knowledge that a department is equipped to carry out large numbers of tests may induce workers, even in other organizations, to make frequent requests for samples to be tested, resulting in an increased workload and the accumulation of masses of data of doubtful value. Occasionally an observant worker using manual methods will encounter some unusual aspect of a test which he has carried out and this may lead to further investigation or even some new discovery. Automated equipment may either miss this completely, give a false reading or report it as an error. A fault may result in the loss of samples and of results which cannot be retrieved.

An important effect of automation is the reaction of staff who in many industries have felt, with some justification, that they were being replaced by machines[2]. To date, this attitude has not been apparent in laboratories but it is quite conceivable that a feeling of insecurity could arise. Some people are not 'machine conscious' and are nervous when working with equipment which they do not understand. Automation should be viewed as a means of producing more results not as a means of reducing staff.

Once automation has been installed and traditional methods abandoned, there is a very real danger of a laboratory becoming totally machine dependent. In the event of a major breakdown, the laboratory may be incapable of producing any results because of the absence of standby facilities or staff experienced in manual methods. This condition may delay important work or fail to produce results needed urgently. Staff using automated techniques should also be trained in traditional methods so that a situation such as this can be overcome.

The decision on whether or not a particular test should be the subject of automation depends on the number of tests to be carried out and on the relative cost of carrying out each test. There is clearly no advantage to be gained from investing large sums in equipment to perform only a few tests, but if considerable numbers are involved the capital outlay spread over many samples represents only a very small cost per sample.

Costing must be accurately carried out and must allow for all expenditure including equipment, reagents, sample vessels, maintenance charges, replacement parts, salaries, etc. The cost per sample can then be calculated and compared with the cost using non-automated techniques.

In recent years, computers have become fairly commonplace in laboratories, mainly in association with automated techniques. With the arrival of solid-state electronics and integrated circuits, they have become more versatile, easier to use, smaller in size and less costly to buy and to operate.

Computers are advanced calculating machines with the ability to store data and to receive and store instructions and with facilities for programming. They are capable of carrying out the most complex calculations at very high speeds, some of which could not possibly be done by any other means. The capability of a computer is referred to in terms of its ability to store and process large amounts of data and to carry out a wide range of calculations. Being a purely electronic device, its action is to convert information into electrical signals which it stores and uses to perform its calculations; it will then convert the signals back into intelligible results. Data and instructions are stored in 'memories' and the greater the capacity of this memory, the greater the amount of information it can handle. It follows then, that each computer has a capacity depending on its design and on its 'language'. The language is the system which is used to enable these calculations to take place and must be programmed into the computer. Some languages are very simple and easily understood by the user, who is thus able to write his own programs; other languages require a considerable knowledge both of mathematics and computer techniques and it is usual to buy programs written in these languages from the manufacturer. In general, the complex languages are more powerful and their use enables the computer to handle more data and instructions than it would be able to do if a simpler language were used.

A computer may operate in a number of different ways; as an advanced high-speed calculator, as a data processor directly coupled to the output from an instrument or as a device to control other equipment. The data storage facilities may be used for record-keeping, stores management, accounting, etc. In some circumstances it may be used in several ways simultaneously; for example, it can be used in conjunction with a scintillation counter and be programmed to change samples automatically, convert raw data into corrected counts (by allowing for background, efficiency and sample quenching), change the settings of the counter to suit different isotopes, reject samples below a predetermined level of activity and display the results in any one of several different ways. It is also possible to interrupt the counting routine to use the computer as a calculator. A program may also be inserted to perform a particular mathematical routine.

Data input may be in the form of a keyboard, punched tape, magnetic tape, etc., and the output can operate an electric typewriter, produce a visual display or energize one of several other systems (*Figure 9.1*). A number of input or

Figure 9.1. Laboratory data processing. Computers collect data from a number of laboratory instruments and process this information to speed up quality control of steel samples taken from furnaces

output terminals can be distributed throughout the organization, all being coupled to the same computer. In some areas, it is possible to rent time on a central computer using only a terminal in the building, thus avoiding the need to purchase one's own computer.

There appears to be no limit to the application of computer technology to laboratory work and it is in this area that science will surely experience many of its most dramatic changes. Laboratory managers are already experiencing difficulty in keeping abreast of new developments in computers and this situation will doubtless prevail for some years. More than ever, equipment is being designed with built-in dedicated microprocessor or computing facilities and it is quite probable that, in the future, all but the simplest devices will incorporate their own data handling and display systems. Even the relatively inexpensive pocket calculators of today are more powerful than many of the large computers of only a few years ago, and they have become cheap enough to be considered as almost disposable items. Today's scientific pocket calculators surpass in most respects the Eniac machine of 1946 which weighed 30 tonnes and consumed several kilowatts of electrical power.

Several relatively inexpensive calculators are now available with standard programs hard-wired into them, and some models can be programmed to perform standard calculations and to print out their results. They are, in fact, evolving so rapidly that it is becoming increasingly difficult to differentiate between the calculator and the computer.

REFERENCES

1. Piggott, R., 'Computers in medicine', *Laboratory Equipment Digest*, 3, 49 (1976)
2. Johnson, A.P., *Organisation and Management of Hospital Laboratories*. Butterworths, London (1969)

FURTHER READING

Malvino, A. and Leach, D., *Digital Principles and Applications*, 2nd edn. McGraw-Hill, New York (1975)
New Scientist, 78, 648–666 (1978)
Rose, M., *Computers, Managers and Society*. Penguin Books, London (1969)

10
Management Techniques and Functions

So far, this book has been concerned with the purely practical aspects of laboratory organization and management. Classical management techniques have only been mentioned in passing to support specific practical points.

Many of those engaged in laboratory work regard 'management' as an esoteric subject that has little to offer towards the solution of everyday problems. They are also suspicious of the jargon used by management specialists. It is the authors' contention that whilst laboratory management should remain an essentially practical subject, much could be done by the application of some management theory to improve the existing standard of laboratory administration. At present, too many laboratories appear to be not only poorly organized but also suffer from such appalling mismanagement that it is difficult to understand how or why they survive at all.

Because a person is a particularly productive or original scientist or technologist, there is no reason to suppose that on promotion he will be transformed into an efficient manager, able to lead the staff of his new department with skill and persuasion. For too long laboratory managers have been appointed primarily on the basis of their scientific expertise and very little thought appears to have been given to their managerial ability, or to their need for management training. On the other hand, in those laboratories where professional managers with little or no scientific training or experience have been appointed to positions of high responsibility, the long-term results have not always been as successful as they might have been. Clearly there is a need for classical management techniques to be integrated with science training so that they may be related to the practical day-to-day life of the laboratory. Several basic techniques are available to enable managers to accomplish their allotted tasks.

Forecasting

To some extent all good management depends on an ability to predict future events. The laboratory manager must use this ability as a means of ascertaining the demands likely to be made on his staff and on his financial and other resources.

The maintenance of accurate and accessible records in a form that may be readily comprehended will enable the manager to observe trends in workloads, etc., so that he will be able to predict the need for an increase or decrease in his work-force. Personnel records, together with the help and advice of the organization's personnel officer, will enable him to prepare for the future loss of senior staff due to retirement, etc., so that other staff may be given appropriate advance training or be otherwise prepared for promotion. Knowledge of current scientific developments will enable him to predict which products are likely to be in greater or lesser demand, or which tests are likely to go out of or come into use. All this information will be of use in predicting future staff requirements.

Similarly, accurate records in respect of the frequency and cost of equipment servicing will enable him to forecast the most advantageous time for replacement.

The prediction of future events from a study of past records is a useful exercise but it is by no means infallible. There is no guarantee that because sales have increased by 10% each year for the past 10 years, there will necessarily be a corresponding increase next year — external factors could produce an unpredicted decrease in demand. At least some of the difficulties currently being experienced by the universities in the UK are due to failure to realize that the post-war expansion could not continue indefinitely. The post-war bulge in the birth-rate has now passed through the universities and the increasing demand for academic staff in industry and public service is by no means sufficient to make up for the rising output of graduates. Without doubt, the secret of accurate forecasting is the gathering together of all the information (or as much of it as possible) with a view to its objective examination in the light of past experience. Foresight is a difficult skill to acquire and success may well owe more to luck than judgement, but nevertheless the key is adequate and accurate information.

Planning

The technique of planning has been defined as the determination of a course of action in order to achieve a direct result. To be effective, a plan needs to set out the methods and resources that are to be used and also to give the expected timing and sequence of events. If management is about getting things done and if valuable resources are not to be wasted, it is evident that planning has an important part to play in effective management. To some extent, the production of a plan to achieve a particular end will limit the range of action that may be taken, in that having been produced, staff will be expected to keep to it. In the entire absence of planning, one is left with a situation in which all decisions are *ad hoc,* and the result is chaos. Unless one is prepared to rely on chance and trust that next year's requirements will be substantially the same as those for the current year, accurate forecasting must be considered an essential prerequisite for effective planning. However, given that future requirements can be forecast with reasonable accuracy, planning consists of relating the resources likely to be available to the anticipated demand. Sometimes the resources available will not match the expected demand; in these circumstances, either the resources (finance, manpower, production capacity, etc.) will have to be increased (or decreased) or the plan will have to be modified. Although, in the interests of proper control, plans must be adhered to they must not be so rigidly defined as to remove all trace of flexibility. Many plans have to be altered from time to time to take account of changing circumstances. The flexibility of planners and their plans is a good measure of their effectiveness.

One of the difficulties frequently encountered in planning is that although recent experience in the field is a decided asset to the planner, he should not be so closely involved in the present work of his department that he is unable to think clearly of the future. He must be able to take a detached view of current departmental problems in order to foresee which long-term improvements would be beneficial.

Organizing

Organizing may be defined as relating a plan to the staff and other resources necessary to carry it out in the most efficient manner. It must be assumed that as all decisions have previously been made at the planning stage, all specified jobs must be done. Organizing is a process of deciding who does what, when it is to be done, how it is best done and what resources are needed to do it.

The techniques involved may be applied both to routine tasks and to projects which only arise infrequently, and also can be adapted to such problems as the assessment of staffing levels, the establishment of the most satisfactory sequence of various activities, the most expedient methods to be used, the allocation of laboratory space and facilities and the allocation of specific tasks compatible with the employees' particular skills and experience. The techniques used are certainly time consuming but, in terms of the resulting increased efficiency, they must be considered as worth while.

Successful organization begins with a critical study of working methods in order to establish ways in which they might be made more efficient. It is essential to interpret the results of work study exercises correctly and to view them in practical terms rather than to become too involved in the theory. The objective must be to improve efficiency and not carry out such an exercise for its own sake.

Work study is the analysis of performance and the study of the information collected in order to determine whether or not the best use is being made of available personnel and resources.

Work measurement is the assessment of the time taken to perform a particular task to a defined standard and forms a basis on which to compare the results with those obtained by using alternative methods (method study). Relative costs of these methods must, of course, be taken into consideration.

Process charts are used to record observations made in order to determine what materials and equipment are used in a given procedure.

Multiple activity charts record the activities of several persons on a common time scale and can draw attention to 'bottlenecks'. If, for example, several workers are preparing samples for analysis using one instrument, delays may be caused if all the samples are ready at the same time. One would have to consider the alternative of staggering the preparation, installing more instruments or automating the process.

Flow diagrams are useful for studying the routes used by staff and materials during the working period. Some relocation of work might reduce wasted time and physical effort. Some years ago, it was not uncommon for all laboratory balances to be housed in one room, which necessitated a great deal of movement of personnel within the building. Improved designs of balances have made the balance room obsolete and the instruments are now accommodated in the laboratories where they are needed.

Activity sampling consists in making observations to establish the precise activities of groups of workers in order to determine what percentage of the working period is spent in carrying out specific tasks. In one laboratory, it was found that a group of people were spending four days each week carrying out analyses and one day calculating the results. The provision of a small computer now enables them to spend five days on analysis, the calculations being made as the work proceeds.

Operational research is a scientific approach to complex problems in which model systems are used to compare the likely outcome of alternative decisions. It is often used as an aid in the determination of policy where a number of alternatives exist and where resources are limited. Its aim is to suggest the most satisfactory solutions to problems of a general nature, such as deciding on staffing levels or finance.

Critical path analysis[1] is an extremely useful technique and may be applied to large and small problems alike, ranging from the initiation of a new technique to the building and equipping of a new laboratory complex. It consists of listing all the jobs which must be done, estimating how long each will take and expressing this information graphically. This is done in order to identify the logical sequence of events and to establish which require the longest time to complete. This sequence is the critical path and indicates the minimum possible time necessary to complete the work. As the work proceeds, any unforeseen delays can be compensated for by adjustment of the less critical time periods and a new

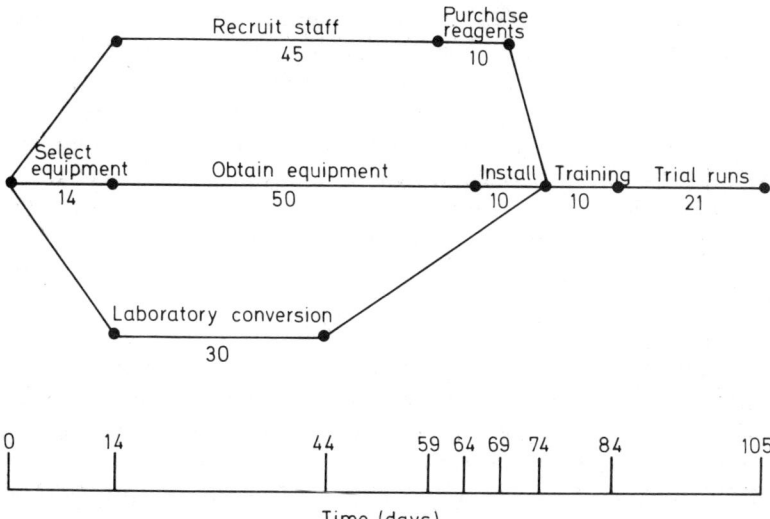

Figure 10.1. A simple example of critical path analysis

completion date can be accurately forecast. *Figure 10.1* shows how critical path analysis may be used as an aid in the establishment of a new laboratory technique. In this example the various tasks are listed, together with the time required for their completion:

Laboratory conversion	30 days
Staff recruitment	45 days
Equipment selection	14 days
Equipment delivery	50 days
Equipment installation	10 days
Reagent purchase	10 days
Staff training	10 days
Trial runs and corrections	21 days

These activities are arranged in logical sequence and the critical path established. Less critical events can be fitted to this path in any convenient way. In the example, it will be seen that a minimum of 105 days is required from the starting date to completion of the project.

The value of graphical presentation of information as a guide to organization should not be underestimated. Uncollated data spread over numerous documents is of little use but its presentation as a graph or chart provides valuable guidance on past performance, demonstrates the present position and possible flaws in the system and simplifies the task of forecasting. It is extremely difficult, for example, to organize work schedules for a number of people without referring to annual leave lists, or to organize a meeting without knowing the dates when participants are likely to be working elsewhere. The use of year planning charts provides instant visual information for such purposes.

The allocation of laboratory space is a continuous organizational problem, particularly in an establishment whose work is of a varying nature. It is not unusual for one department to have a genuine need for increased space, whilst another can be contracted to some degree without detriment to efficiency. Such reorganization almost invariably requires alterations to the laboratories and if the interruption of work schedules is to be minimized a time table must be prepared and adhered to rigidly. Events such as open days and exhibitions can tax the laboratory manager's organizational abilities to the fullest extent and he may have to resort to a count-down system, in which certain tasks must be completed by predetermined dates.

The unexpected absence of staff because of sickness or other causes frequently necessitates the reorganization of planned tasks in order to provide the necessary labour to carry out essential work. Situations of this type usually arise at short notice and it is sound policy to prepare prearranged plans so that these may be put into operation immediately.

The organization of the routine work of the laboratory should be updated at regular intervals, as dictated by changing circumstances. Working methods which are acceptable today are not necessarily the most satisfactory for all time and any change which will result in improved efficiency is worthy of consideration. The cost of labour is high and must be taken into account, together with all other factors, when deciding how and when a particular job should be carried out.

Motivating

The most valuable asset possessed by any establishment is its staff and it is surprising, therefore, that whilst many senior workers are conversant with the intricacies of their equipment, they may have little or no concept of the mental processes which inspire people in their work. The cost of staff accounts for about 65% of the running costs of the average laboratory; this is a very considerable proportion of the budget to allocate for any single item. On the assumption that the staff have been selected with care, it would not seem unreasonable to expect them to be employed in the most advantageous way possible. Much has been written on the subject of human behaviour and psychology and it is not the intention here to duplicate these discussions. It is, however, useful to consider the essentials of the subject in order to understand what factors stimulate individuals in their daily tasks and what problems may result in dissatisfaction and loss of enthusiasm.

People work for a variety of reasons and these assume different priorities in individual cases. There is an obvious need to survive and an income is therefore a necessity. It is easy to assume that pay is the main, if not the sole, consideration but the evidence does not support this cynical assumption. The opportunity to increase income by accepting promotion or by changing jobs is often rejected by staff who prefer to continue with their present jobs which provide them with satisfaction that they are not prepared to sacrifice. Job satisfaction, although difficult to define, is a very real experience and is achieved in a number of ways. For some it is a feeling of doing a worthwhile job, or of working to a high standard, whilst to others it gives a sense of achievement or of reaching a particular status which earns them the respect of their fellows, or an intense interest in some aspect of the work. Man needs to feel that he is using his skills to advantage and if these are skills not possessed by his associates, he acquires a certain 'one-upmanship' which satisfies his ego. Some find satisfaction in responsibility and status which enable them to put something more into the job above what is normally expected of them.

It is only possible to generalize to a limited extent in the consideration of aspects of human behaviour and motivation. People are individuals and have differing needs and indeed these will vary in an individual as circumstances change. In a young person, the driving force may be competitive, perhaps in the hope of being promoted in advance of his colleagues, but as he becomes more mature and acquires family responsibilities, his priority may be in terms of salary improvements. As retirement age approaches, he may find satisfaction in work which requires the benefit of his experience more than physical effort. It is essential for the manager to consider the needs and aspirations of each individual as far as possible, because only then will he be able to fit staff into jobs which provide the greatest fulfilment.

In the laboratory, people are part of working groups known as primary and secondary groups. A primary group consists of a small number of people who are mutually associated because of similar status, close physical contact or common aspects of their work. Secondary groups are those which exist to do a particular job, such as the entire staff of a factory or department. Members of primary groups tend to interact in ways which enable them to assign certain roles to each other. In secondary groups, these roles are usually defined by rank and by the particular duties of the members. The relationship of any individual to the rest of the group will depend on his conforming to certain standards and the support he gives by carrying out his work in a way which is acceptable to the other members. The failure of one individual may result in the failure of the entire group to achieve its objectives.

Personality clashes in a working group may result in diminished output or poor quality of work, even though the individual members are of a high calibre. It may at times be advantageous to transfer staff from one working unit to another in order to avoid situations of this kind.

A reasonable level of staff turnover can be expected and is in fact good for both the organization and the individuals. Experience and exchange of ideas are highly desirable and changing jobs is, in most cases, the only way to progress in one's career. Should turnover reach unacceptable proportions, the manager must investigate the situation to determine the cause. Almost certainly there will be some defect which results in dissatisfaction among the staff. There may also be a high level of absenteeism and persistent sickness of an ill-defined nature, both of which are symptomatic of the problem.

Dissatisfaction in a particular work situation may be brought about by a number of causes. Inadequate leadership can result in failure to reach targets, giving rise to a sense of frustration just as acute as that resulting from trying to work without the proper equipment and materials. Working under supervisors who make unreasonable demands, who criticize unjustly, or who fail to offer a word of praise when a job has been well done, frequently results in a feeling of resentment.

There is little doubt that most people perform best under slight pressure, both in terms of quantity of work and level of attainment. Some challenge should be present in order to stimulate a desire to overcome problems and so to experience a sense of achievement. If, however, this challenge is persistently too great for an individual, he may fail to reach his goal on numerous occasions and hence develop a feeling of inadequacy to such an extent that he will give up the struggle.

An efficient working team is not assembled overnight but takes a considerable amount of effort and expertise to develop. In general terms, one has to accept a compromise between the needs of the organization and the ability of the team to carry out its prescribed tasks.

Co-ordinating

Every member of the staff, and therefore each group, makes a contribution to the joint efforts of the organization, but however great this may be, much of it can be wasted and work may be duplicated if their endeavours are not co-ordinated. Most workers will have experienced situations in which they have carried out some arduous task only to find that a colleague has done the same thing. The larger the organization, the more likely it is that something of this nature will occur. Similarly, a task may be overlooked on the assumption that it has been done by somebody else. An employee may be unable to work without a particular piece of equipment, unaware that another is available elsewhere in the building. The co-ordination of effort requires particular attention when work is being carried out jointly between a number of separate organizations.

Many of these problems result from poor communication and unsatisfactory integration of resources. Communicating as a management technique is discussed later in this chapter and its value as a factor in co-ordination cannot be overemphasized. Interdepartmental discussions and seminars are convenient ways of informing staff of the work which is being done by others in addition to the facilities which are available. There would, for example, be no point in one department setting up facilities for carrying out a few chlorine estimations if another department was already providing this service on a routine basis. Surplus equipment can be offered to units who may need it and employees not required by one department can sometimes be transferred to another, possibly on a temporary basis, rather than recruiting more staff. The work of a laboratory which is being refurbished may be temporarily accommodated in another department, thus enabling work to continue with the minimum of interruption.

Successful co-ordination is dependent on the manager having access to all necessary information and on his discussing the problem with all those involved. He may well be the only person in the establishment who is fully aware of the wide-ranging factors, such as finance, staff, equipment and other resources available.

Controlling

The technique of controlling may be described as the process of comparing actual performance with planned performance. It is a continuous process that needs to be applied with sensitivity to the work, so that the appropriate remedial action can be taken. Where production work is concerned, output may have to be accelerated to reach previously agreed targets and occasionally slowed down to avoid overproduction and the resultant storage and financial problems.

The amount of correction to be applied is important; over-correction may be just as embarrassing as no correction at all. Violent swings are generally best avoided by regular small adjustments rather than by large adjustments infrequently applied.

The most common and probably the most important application of the control process in the laboratory is the attention which is given to global and specific budgets to ensure that they are neither under- nor over-spent. Budgets are best illustrated graphically so that their warning signs may be interpreted at a glance. Some budgets, such as those concerned with wages and salaries, can be fairly evenly allocated throughout the financial year and, provided that not more than one-twelfth of the annual budget is expended each month, all will be well. Other budgets are used in a less regular fashion. Those for non-recurrent expenditure such as major capital equipment tend to expend more than half of the allowance in the first few months of the year, whilst those for consumable items reach a peak in certain specific months. For research laboratories, this peak tends to occur in October when all the staff have returned from their holidays, whilst for schools, the peak expenditure occurs just before the major examinations in May—June.

If the cumulative total expenditure is plotted each month on a graph which shows a straight-line cumulative target expenditure for the year, it is possible to determine quickly and precisely by how much the target has been exceeded (or otherwise). Past experience will show how much deviation above or below the optimum may be tolerated for any given month.

Communicating

Communicating is essentially a process which provides the link between staff at all levels and which enables information, instructions and decisions to be conveyed from their point of origin to all those who need to be informed. It is without doubt one of the most important functions of managers and in the absence of a system of adequate and efficient communication, management cannot fulfil its role.

From time to time, managers have been accused of withholding information from employees regarding matters which have a profound effect on their employment and working conditions. Occasionally this may be part of a deliberate policy but is more often due to poor communication which, if improved, could avoid or at least substantially reduce much of today's industrial strife. At the other extreme, submerging staff with a plethora of irrelevant and useless information can be equally demoralizing and damaging to good relations.

Mention has been made earlier in this chapter of working groups. It must be emphasized that unless each member of a group is fully aware of what his colleagues are doing, the group cannot function properly. It is necessary to be able to communicate upwards to senior management, sideways to colleagues and downwards to subordinate staff. Communications may be one way, and such is the case when issuing instructions or giving information on a particular subject, or they may be two way, e.g. when conveying messages which require comment or reply. It is vitally important that two-way communication should reach a conclusion and if a reply is called for it must be given. Enquiries must be followed through and answers provided.

Direct communication from one person to another is the simplest and quickest way of passing information or instructions and it is usually verbal — either face-to-face or by telephone. Unfortunately it is not always reliable, particularly in a busy organization, as there is a risk of messages being forgotten before the necessary action has been taken. It is usual to follow up verbal messages with written confirmation unless only trivial matters are involved.

Indirect communications usually originate from one person only and they are intended to be conveyed to a number of persons, e.g. a statement of policy directed to all employees. Such a statement may emanate from top management levels and be routed through middle management with instructions to notify all staff.

Formal communications usually take the form of orders, instructions and statements on which management expects action. They are usually in writing and should, if they have been correctly presented, avoid the possibility of ambiguity and misinterpretation. Written memoranda are also used to set down for the record that certain events and discussions have taken place, that specific decisions have been made or that certain instructions were given at a particular time. They also provide documentary evidence in the event of a disagreement at a later date. Despite the obvious usefulness of written communications, the fact that they have limitations should be appreciated. They are a one-way process and cannot easily be altered to take account of subsequent discussions with the recipient. Redistribution of updated communications can be a costly and time-consuming operation which may give rise to possible accusations of vacillation.

Informal communications are usually conveyed verbally, a method which allows for a degree of consultation to take place on the spot with the effect that messages may be modified or even withdrawn as a result of such discussions. This process of face-to-face discussion is essential in the building up of team spirit and in the development of mutual respect for one another's views. Face-to-face discussion can help the subordinate to appreciate some of the problems of management and, more importantly, it will give him a sense of belonging and of his personal value to the organization.

Instructions and orders can be graded in terms of their authority; clearly an instruction from a managing director to a very junior employee will need to be accorded rather more consideration than a request from a person of equal rank. Similarly, a reply from the junior employee to the managing director would be couched in more courteous terms than those he would use when acknowledging a request from his colleague.

A direct order, unambiguously given by a senior, demands that it must be obeyed. It may offend the sensitive but it is usually necessary if an emergency situation should develop. In this form, it may shake a slack worker out of his lethargy and avoid the necessity of taking firmer action at a later date. The implied order will usually meet with the best response only from a reliable and dependable worker and is not likely to have the required impact on the inexperienced, unreliable or lazy worker. An informal request will often achieve better results with willing workers and will not offend, although the belligerent person may choose to ignore it until further pressure is put on him.

Numerous communication systems are available, each with

its particular merits in different circumstances. The telephone, once considered a status symbol for the higher executive, is now accepted as an essential item for all members of staff who need to communicate rapidly. A considerable amount of laboratory work requires contact with others outside the organization, such as suppliers, service agencies and workers in other laboratories. An internal telephone system saves endless wasted time by making it possible to contact individuals rapidly. Unfortunately, telephones are frequently abused and it is important to ensure that they are not used for unnecessary personal messages and that conversations are kept as short as possible. Many organizations provide pay phones at strategic points in the building for staff use.

The telex system is a means of conveying printed messages through the telephone network. In the UK it is becoming more widely used to transmit all kinds of messages where speed and accuracy are essential. The message is typed into the keyboard of an instrument resembling a typewriter and is simultaneously reproduced on the recipient subscriber's instrument. Messages may be commented on or acknowledged as long as the line remains open. They may also be transmitted out of working hours when the equipment is left in the 'standby' mode. Although the system is relatively expensive, the extra cost may well be recovered in the increased efficiency and speed of communication.

Rapid communication with overseas contacts may be achieved by the international cable systems. It is normally possible to contact persons in major towns anywhere in the world and to receive a reply within a few hours.

Written memoranda have been mentioned as a method of conveying information and they are normally despatched through the internal post system within the organization. Where the subject matter is important, carbon copies should be retained and filed for future reference. It is sometimes necessary to circularize groups of people with particular items of information; for example, some may need to be sent to selected individuals such as heads of departments and some to all members of staff, whilst others will be for restricted circulation to one or two individuals. This task is simplified if circulation lists are kept, giving the names of those persons who constitute the various groups.

Notice boards are a convenient way of conveying information to large numbers of people. They are not, however, a totally reliable medium, because a surprising number of people seldom study them. Whilst it is sometimes helpful if important notices are printed on brightly coloured paper, all redundant notices should be removed at least once a week. The boards must, of course, be sited in areas where staff frequently circulate.

Information which must be sent to all members of staff may be included in wage packets or salary notification envelopes. This method ensures total circulation without extra cost and with the minimum of effort.

House journals and newsletters are excellent methods for the dissemination of information of a general nature and they are likely to be read in detail if written in an informal and chatty style. They also help to keep staff in touch with the varied activities of their colleagues in other departments.

Group meetings and seminars which are arranged to discuss particular topics or aspects of the work are not difficult to organize and they certainly provide a satisfactory platform for the exchange of ideas and for airing differences of opinion. They should not take place too frequently and adequate time must be available for replies and discussions. They provide excellent two-way contact between management and staff and can result in a drastic reduction of written communication between the two. Care must be exercised in selecting the time and venue for the meeting and the opening speaker must be chosen with caution. Too often he conveys a sense of 'bringing the tablets down from the mountain' and intimidating those who may question his views. The discussion may be monopolized by the stronger characters among the audience who have confirmed points of view, a situation which reduces the value of the meeting for the perhaps quieter but no less aware members of the staff. Nevertheless, informal meetings do have a most important part to play in bringing about a better understanding, reducing disagreements or misunderstandings and improving co-operation between working groups and individuals. The scope is widened further if workers from other laboratories are invited to attend, either as speakers or as participants in the discussion.

Communication between scientists and technologists and

their colleagues in other organizations frequently takes the form of papers which are published in the appropriate journals. The subject matter may be the proposal of new theories, the exchange of ideas, the reporting of experimental results or the introduction of new techniques. Numerous journals exist for this purpose, some being of a general scientific nature whilst others are concerned with highly specialized subjects ranging from astronomy to zoology. Scientific libraries subscribe to a number of such journals, covering the disciplines appropriate to the particular organization. Many journals can be obtained on loan from other libraries or, alternatively, reprints of articles may be obtained on request from the authors. When an article is submitted to a journal, the editorial staff will decide whether or not it is suitable for publication, or may suggest alterations or may even reject the article. Some journals provide a number of reprints free and make a charge for further copies.

The volume of published papers is so enormous that without some form of abstracting service it is impossible to keep abreast of current developments on a worldwide basis. Several computerized information services exist in numerous countries and subscribers are kept up to date on published work in specified areas of interest.

There is a tendency to judge scientific and technical expertise by the number of papers published by individuals, with perhaps insufficient regard for the quality of the work which they represent. This is a regrettable situation but it is difficult to suggest an alternative standard which could be used, particularly where original research work is concerned. The informed reader will, of course, arrive at his own opinion of their true value but junior workers often find that they are under some pressure to publish at regular intervals in order to further their careers.

It is occasionally desirable to inform the general public, or groups of interested persons, of the main lines of work of a scientific establishment. One successful way of achieving this is to hold open days. Organizations which depend on outside sources of finance also have an obligation to invite donors to the laboratories to see for themselves what use is being made of their money. Although open days invariably require considerable effort on the part of the staff, they are good for public relations and can be very rewarding.

The preparation of displays must be started well in advance, due allowance being made for the continuation of work which cannot be interrupted or postponed. A carefully prepared count-down is of great value in ensuring that preparations are completed in time. The overall impression is greatly enhanced if display work is standardized as far as possible by the use of uniform styles and sizes of lettering and artwork. Each display should be complete in itself and should explain to the layman exactly what is happening in simple terms.

It is not feasible to have all the laboratories on display, for a number of reasons. Some equipment and materials must be removed from the rooms which are used for exhibitions and other rooms will be needed to house these items. Laboratories may also be needed for on-going work.

A guest list must be prepared and formal invitations must be sent to visitors. Arrangements should be made for their acknowledgement so that the organizers will know how many people propose to attend. It is important to treat visitors as guests of the organization and to arrange for them to be greeted on arrival and also provided with lists of exhibits and directions on how to find their way round the building. Staff must be detailed to conduct visitors and direction notices should be posted at strategic points. Employees whose work is on display must be available to explain it and to answer any questions asked by visitors. Due consideration must be given to visitors' safety and they must not be exposed to any danger. When visitors have left, a thorough check of the building must be made to ensure that all equipment and supplies have been switched off and that no fire hazards exist from such sources as cigarette ends. Some of the display work may be considered suitable for further use and should be stored for this purpose. All other material should be disposed of and the laboratories restored to normal working conditions as soon as possible.

In any communication process, the need for clarity and brevity is paramount. In a lengthy and verbose document, the essential point of the message may become obscured by a fog of irrelevant statements; so one would do well to follow the advice of William J. Mayo (1861–1939): 'Begin with an arresting sentence, close with a strong summary; in between, speak simply, clearly and always to the point; above all be brief.'

REFERENCE

1. Lockyer, K.G., *An Introduction to Critical Path Analysis*. Pitman, London (1970)

FURTHER READING

Argenti, J., *Management Techniques, a Practical Guide*. Allen & Unwin, London (1969)
Burns, T. and Stalker, G.M., *The Management of Innovation*. Tavistock Publications, London (1971)
Vroom, V.H. and Deci, E.L., *Management and Motivation*. Penguin Books, London (1970)

Appendix 1
Certificate in Medical Laboratory Management

This appendix gives the regulations and study scheme for courses leading to the Certificate in Medical Laboratory Management of the Institute of Medical Laboratory Sciences.

Regulations

The Certificate in Medical Laboratory Management can be obtained by passing examinations at the satisfactory conclusion of an approved course, occupying at least 270 hours, and by the assessment of project work. Candidates for entry to the examination leading to the certificate must be either

(1) Fellows or Associates of the institute, or
(2) Senior members of overseas institutes or associations having objects similar to those of the institute, or of other relevant professional bodies approved for the purpose by the institute.

Candidates must be sponsored by establishments for further education which must satisfy the governing council that each candidate has attended for not less than two-thirds of the prescribed time in each stage of the course; candidates will be required to satisfy the examiners in two written papers, each of approximately three hours, and separately in a project.

By 1 April in the year following the commencement of the course, the title of the proposed project must have been submitted for approval by the council. The project title should be amplified to indicate that the project will have relevance to management in the context of medical laboratory sciences and that the report on it will demonstrate some originality of thought and will draw conclusions from the investigation to be undertaken. The project work and the report on it, which should be of between 3000 and 6000 words, shall be submitted at the conclusion of the course and must conform with the requirements specified under 'Projects' on page 11 of the Regulations for Membership and Entry to Examinations booklet.

Courses which have been approved by the institute may be offered by colleges or polytechnics in the UK. When making application for approval of a course the college or polytechnic must

(1) Satisfy the institute that the course will occupy approximately 270 hours distributed over two years;
(2) Appoint a board of studies, the composition of which shall include at least two Fellows of the institute; and
(3) Submit at least six months before the proposed date of commencement of the course
    (a) a curriculum, including details of time allocated to the major sections of the course, and the proportions of time to be devoted to formal lectures, projects and case studies;
    (b) a detailed teaching scheme, based on the study scheme set out below;
    (c) the names, professional qualifications and appointments of members of the full-time and visiting teaching staff employed for the course;
    (d) the names, professional qualifications and appointments of members of the board of studies appointed to advise on the course; and
    (e) details of the proposed arrangements for project work in the application of management techniques.

Study scheme for management of scientific departments

This course is designed for staff who may be or have been appointed to posts with managerial responsibility in scientific departments. The intention is to inculcate a sophisticated approach to management and an awareness of the management techniques which modern practice demands. As successful completion of the course should assist in the development of individual potential, participatory methods, including projects and case studies, must be incorporated as fully as possible.

STRUCTURE AND FINANCE OF THE HEALTH SERVICES

Organization structure of the NHS and related scientific services.

Allocation and employment of capital within the Health Service and Social Services.

PRINCIPLES AND PRACTICE OF MANAGEMENT

Purpose and nature of management.

Accountability, authority and responsibility; delegation of authority and span of control.

Role of the manager within the organization.

Personnel

Man management, job analysis, job description, discipline, laboratory safety.

Interviewing techniques associated with: selection, induction, counsel, discipline and grievance handling.

Personnel assessment and evaluation.

Training and education of staff, in-service training, evaluation and validation of training processes.

Role of the various bodies involved in education and training (e.g. Institute of Medical Laboratory Services, colleges, Council for Professions Supplementary to Medicine, national certificate joint committees). Technician Education Council and Scottish Technical Education Council.

Roles of personnel departments, and relationship with these departments.

ORGANIZATIONAL SOCIOLOGY

Social systems in the NHS at national and district levels and their effect on scientific services.

Functions of, and distinctions between, the following groups: professional and managerial; advisory and executive those with power and those with authority.

SOCIAL PSYCHOLOGY

Individual behaviour; motivation and attitudes.

The learning process and training.

Influencing individual behaviour; leadership; discipline; incentives and resistance.

Group relationships, primary and secondary groups, conflict, co-operation and objectives.

Special problems related to: highly automated and mass-production laboratories; the very small hospital; the large complex teaching hospital and group departments.

LEGAL ASPECTS

A summary of the sources of law, branches of law and law courts.

Knowledge of the law relevant to managers and in particular managers in the scientific service, including: the common law of contract; negligence; relevant legislation on employment, environment, labour relations, safety, and the use and misuse of drugs and medicinal products.

LABOUR RELATIONS

An appreciation of the different perspectives in which labour relations are seen by national bodies representing, respectively, employees, employers and professional groups.

The parties in industrial relations: national and local. National disputes, local disputes (including discipline, grievance handling and appeal procedures).

Trade unions

Purpose, organization and legal status.

Whiteley Council

Purpose and composition; mode of action.

MANAGEMENT TECHNIQUES

Financial control

The laboratory manager and capital utilization. Elements of costing and cost analysis. Theory and practice of budgetary control and output budgeting.

Planning – short, medium and long term

Internal and external environment of the department. Planning objectives, criteria and constraints for regions, areas, districts, departments and individuals. Methods of assessing results.

Communication

Purpose and principles. Effective transfer of information, ideas and values to achieve organizational goals.

Methods of communication

Channels of communication, official and unofficial. Organization, interpretation and presentation of facts and ideas. Convening and organizing meetings and conferences. Role of the chairman. Agendas and minutes. Effect of organization structure on communications. Barriers to effective communication.

Management services

Design and layout of department. Creating an awareness of the value of management services and their applications in the scientific department.

Stock control methods

Statistical control. Charts, sampling theory, probability theory.

Management information

Systems and uses. Record systems. Documentation and reporting. Basic understanding of computers. A systems approach to organization and management.

The authors are most grateful for permission to reproduce the above information. Further details may be obtained from the Institute of Medical Laboratory Sciences, 12 Queen Anne Street, London, W1M 0AU.

Appendix 2
Diploma in Laboratory Management

This appendix gives the regulations and syllabus for the Diploma in Laboratory Management of the Institute of Science Technology.

Introduction

The syllabus has been prepared to provide a scheme of education, training and examination for senior technicians involved in the management and administration of laboratories.

Examination

The examination is open only to members of the institute and the full regulations governing the examination are obtainable from the General Office, Institute of Science Technology, 345 Gray's Inn Road, London, WC1X 8PX.

The examination will consist of:

(i) Two written papers each of three hours' duration:
Paper 1. Laboratory Procedures and Organization
Paper 2. Laboratory Management.
(ii) (a) A project based on laboratory management: the candidate will be expected to submit a detailed record of the project.
(b) Candidates will be expected to submit a title and outline of their proposed project for approval by the

examiners. This should be enclosed with the application form.
(iii) Candidates may be called for oral examination.

Courses of study

The syllabus has been devised on the assumption that candidates who have followed a recognized course will have completed a minimum of 480 hours of study.

Any establishment wishing to introduce a course leading to a diploma examination must inform the institute prior to commencing the course.

Prize

An institute prize may be awarded annually to the outstanding candidate in the diploma examinations.

Syllabus

A. LABORATORY PROCEDURES AND ORGANIZATION

(a) *Fittings and services*
Flexible and fixed permanent furniture and equipment of laboratories, dark rooms, workshops and stores — principal factors governing layout and arrangement, including benches and workplaces, essential services and supplies, ventilation and lighting.

(b) *Receipt and despatch of goods*
Purchase, issue and receipt of materials and equipment including alcohols and poisons, apparatus, raw materials and samples. Rail, post and road service requirements. Unpacking and checking of goods and samples. Regulations governing imports and the legal liabilities involved.

(c) *Storage of materials*
Storage and preservation of apparatus and materials, materials requiring special care including flammable,

labile, poisonous, highly reactive materials and compressed gases.

(d) *General maintenance of the laboratory*
Services, apparatus, stores, equipment and furniture. Periodic inspection, repair and renewal.

(e) *Laboratory safety*
Precautions against fire and explosion, electrical hazards, gas supply and storage risks, corrosives and poisons, dangers from glassware and cutting tools, precautions against infection. Correct action in emergencies, location of main switches and valves, fire extinguishers and fire-fighting equipment. First-aid treatment of casualties from the most likely forms of laboratory accident.

(f) *Radiological procedures and safety*
The nucleus, isotopes, radioactive decay and half-life, interaction of radiations with matter and with living organisms, film badges, pocket dosimeters and simple monitors. Principles of protection and precautions against ionizing radiations from sealed and unsealed sources. Control of contamination. Handling and storage of radioactive substances, codes of practice, rules governing the disposal of waste.

(g) *Office work in relation to the laboratory*
Laboratory records including ordering procedures, budgeting, petty cash, stock and material control, inventories. The collecting and indexing of technical information, including diagrams, slides, film strips and films. The techniques of document reproduction.

(h) *Sources of information*
Reference books, standard textbooks and periodicals, data and technical information and catalogues.

(i) *Demonstrations and exhibitions*
Methods of displaying apparatus and processes, charts and diagrams.

B. LABORATORY MANAGEMENT

(a) *Management*
Nature and process of management, basic principles involved. Organization structures of research and

educational establishments. Functions of the laboratory manager — planning, organizing, controlling, co-ordinating, communicating and motivating.

(b) *Human relations and personnel*
Work groups; autonomy, satisfaction, decision-making, automation. Social groups at work. Industrial psychology and morale. Industrial relations, economic, social and political aspects. Trade unions and the TUC. Employers' and employees' associations. Educational, research and professional bodies. Job analysis, description, evaluation and grading. Job design. Recruitment and selection of staff. Career development, personnel assessment. Training and education of staff, role of various bodies involved (e.g. IST, colleges, TEC). Methods of training, measurement of training success. Relationship with personnel and administrative departments. Styles of leadership. Laboratory discipline. Staff welfare, sickness, personnel counselling, absenteeism.

(c) *Work planning*
Working to objectives, work policies, allocation of resources. Work study and measurement, methods and practice. Planning and control techniques, including critical path analysis. Trends in organization and policies. Medium and long-term planning.

(d) *Communications*
Principles and practice. Channels of communication, official and unofficial. Written communication, letter, report and publication writing. Laboratory managers' responsibility for communication. Interviewing, public relations, organization of meetings, conferences. Open days. Arrangements of lectures, demonstrations. Reprographics, mass communication techniques, choice of appropriate media. Committee and sub-committees and their role with respect to line management. Committee terms of reference, reports and minutes, role of officers.

(e) *Quality control*
Control charts, sampling theory, probability. Principles of experimental design.

(f) *Financial control*
The laboratory manager and capital utilization. Elements of costing and cost analysis. Capital and revenue expenditure. Budgetary control.

(g) *Computer appreciation*
Electronic data processing — input and output media and devices. Introduction to programming. Application to laboratory technology and management, e.g. stock control and record-keeping.

(h) *Legal aspects*
Contracts: contracts of employment, redundancy and insurance, injury at work and compulsory insurance, common law. Laws of copyright and patent. Statutory Law:
   The Industrial Relations Act 1971
   The Industrial Training Act 1964
   The Offices, Shops and Railway Premises Act 1963
   The Factories Act 1961
   The Fire Precautions Act 1971
   The Health and Safety at Work etc. Act 1974
   The Cruelty to Animals Act 1876
   The Rabies Act 1974
   The Employment Protection Act 1975
Negligence and liability. Responsibilities of employer and employee. Consumer protection. The Sale of Goods Act 1893. Implied statutory terms and exclusion clauses. The Trade Descriptions Act 1968.

(i) *Science laboratory administration*
Planning: Design factors, cost, adaptability, regulations, safety, ease of operation and efficiency for required purpose; sources of information and advice. Planning of new buildings and modifications of old. Organization of suitable services, fixtures and fittings. Specialized rooms. Requirements of industrial, research, hospital and teaching laboratories.
Laboratory record systems: Including ordering procedures, budgeting, stock and material control, inventories, radioactive substances, animal experiments, etc. Information retrieval systems and uses.
Legal requirements and codes of practice: In the handling of poisons, alcohols, radioactive materials and X-radiation, pathogens, biological hazards, quarantine and experiments on animals. Regulations concerning import, export and excise duties. Postal and rail & road regulations.
Maintenance of laboratories, stores, etc.: Provision for

routine care of services and equipment, including safety services and checks.

Safety: Responsibilities under statutes. Role of laboratory manager. Safety organization and procedures. Prevention of risk of injury, infection, poisoning and fire.

The authors are most grateful for permission to reproduce the above information. Details of the various grades of membership and further details of the diploma examinations may be obtained from the Institute of Science Technology, 345 Gray's Inn Road, London, WC1X 8PX.

Appendix 3
Technician Education Council Higher Awards in Laboratory Science and Administration

TEC policy is not to publish syllabuses *per se* but to issue programmes of study based on a system of 'units'. Those relating to laboratory management are: (1) Laboratory Administration IV (TEC U78/512), and (2) Supervisory Studies – Science (TEC U78/513). The former has a value of 1½ units and the design length is 90 hours, while the latter comprises 1 unit with a design length of 60 hours.

TEC U78/512
Laboratory Administration IV

UNIT CONTENT

The unit topic areas and the general and specific objectives are set out below, the unit topic areas being prefixed by a capital letter, the general objectives by a non-decimal number, the specific objectives by a decimal number. *The general objectives give the teaching goals and the specific objectives the means by which the student demonstrates his attainment of them.* Teaching staff should design the learning process to meet the general objectives. The objectives are not intended to be in a particular teaching sequence and do not specify teaching method but, for example, practical work could be the most appropriate teaching method for the achievement of the objectives.

All the objectives should be understood to be prefixed by the words: The expected learning outcome is that the student: —

A   LABORATORY DESIGN

1   Appreciates the role of the supervisor in laboratory layout and design.
     1.1   Explains the relative roles of management and supervisory technician in the design of laboratories.
     1.2   Explains the relationship between function and design for general and special areas in laboratories.
     1.3   Discusses the factors to be considered in the provision and layout of fixed and mobile furnishings of laboratories and special areas such as dark rooms, instrument rooms, stores, etc.
     1.4   Discusses information required by architects, engineers and craftsmen, relating to design, construction or modification.
     1.5   States information required from architects, engineers and craftsmen relating to 1.4.
     1.6   Evaluates information specified in 1.5.
     1.7   Discusses the factors to be considered in the provision, design and layout of laboratory services, e.g. gas, water, electrical power, drainage, waste disposal, heating, ventilation and lighting.
     1.8   Discusses the considerations regarding provision of specialist services (vacuum, steam, compressed gas supplies, etc.)
     1.9   Prepares sketch plans and accompanying explanatory notes both for new buildings and for modifications to existing premises.

B SELECTION AND SUPPLY OF EQUIPMENT

2 Understands the factors involved in the selection and supply of laboratory equipment.
  2.1 Discusses the role of the supervisory technician in the choice of equipment.
  2.2 Explains the factors to be considered in the selection of equipment, e.g. cost, availability, performance, durability.
  2.3 Explains the special factors to be considered in the selection of major items of equipment, such as location, provision of services, installation and commissioning, servicing, running costs.
  2.4 Discusses considerations regarding the purchase and storage of consumable materials.

C MAINTENANCE

3 Understands the principles of maintenance of laboratories and equipment.
  3.1 Explains the role of the supervisory technician in the preparation of schemes for routine maintenance, including liaison with craftsmen and service engineers.
  3.2 Prepares schemes for maintenance of working surfaces, services, furniture (fume cupboards, refrigerators, etc.) and equipment.
  3.3 Evaluates methods for dispersing laboratory reagents to the user.
  3.4 Describes the role of the supervisory technician in the provision and care of living materials.

D GENERAL ADMINISTRATION

4 Applies general procedures of laboratory administration.
  4.1 Prepares a budgetary monitoring scheme given circumstances in which it is to operate.
  4.2 Prepares schemes for the operation of a petty cash account and an imprest account system.

4.3 Describes methods to ensure the correct flow of materials to the laboratory worker, including ordering and stock control procedures.
4.4 Prepares schemes for the usage of laboratories and special areas in accordance with the needs of timetables and work schedules.
4.5 Prepares a booking scheme for specialised equipment (e.g. AVA, mass spectrometer).
4.6 Describes the functions of inventories.
4.7 Prepares an inventory for a given situation.

E COMMUNICATIONS

5 Understands the requirements and methods for the dissemination of information.
   5.1 Lists various types of audio-visual and reprographic equipment.
   5.2 Explains the factors to be considered in the provision, use and control of the equipment in 5.1.
   5.3 Describes and explains the role of the supervisory technician in the organisation of meetings, exhibitions, open days, and practical sessions.
   5.4 Discusses the special needs of lecture-demonstrations.
   5.5 Prepares and presents technical and advisory reports.
   5.6 Advises others in the preparation of reports in 5.5.

F STATUTORY PROVISIONS REGARDING MATERIALS AND EQUIPMENT

6 Understands and applies statutory provisions regarding materials and equipment.
   6.1 Explains implications for the supervisory technician of the regulations concerning import and excise duties on equipment, purchase and storage of alcohol, petroleum spirits, etc.
   6.2 Describes the requirements of the laws relating to experiments on, and the import of, living materials.
   6.3 Describes the implications for the supervisory technician of the laws regarding consumer protection such as the Sale of Goods Act 1893, the Trade

Descriptions Act 1968, the implied statutory terms, and exclusion clauses.
6.4 Explains the implications of laws of contract in regard to purchasing and servicing of equipment.

G SAFETY

7 Understands the role of the supervisory technician in the establishment and maintenance of a safe working environment.
7.1 Discusses the responsibility of the supervisory technician in relation to prevention of injury, poisoning, damage by fire, etc., under the terms of the relevant Acts of Parliament.
7.2 Evaluates the relative responsibilities of management and supervisory technician for the provision and use of protective clothing and safety equipment.
7.3 Describes the role of the supervisory technician in the implementation of regulations concerning the following:
(a) poisons;
(b) radioactive materials;
(c) ionising radiations (X-rays, sealed sources, UV);
(d) non-ionising radiations (e.g. lasers, microwaves).
7.4 Prepares examples of 'good housekeeping' codes of conduct for areas selected from 7.3 (a) to (d).
7.5 Prepares schedules of safe working procedures, and safety check lists.

TEC U78/513
Supervisory Studies — Science

UNIT CONTENT

The unit topic areas and the general and specific objectives are set out below, the unit topic areas being prefixed by a capital letter, the general objectives by a non-decimal number, the specific objectives by a decimal number. *The general objectives give the teaching goals and the specific objectives the means by which the student demonstrates his attainment* of them. Teaching staff should design the learning process to meet the general objectives. The objectives are not intended to be in a particular teaching sequence and do not specify teaching method but, for example, practical work could be the most appropriate teaching method for the achievement of the objectives.

All the objectives should be understood to be prefixed by the words: The expected learning outcome is that the student:—

A RECRUITMENT AND SELECTION

1 Understands the principles used in recruitment, selection of personnel and termination of employment.
   1.1 States the implications for the supervisory technician of the law relating to contracts of employment and conditions of service. (Employment protection, sex discrimination, T.U. and Labour Relations, Equal Pay, Race Relations, Redundancy, Pensions Acts.)
   1.2 Drafts a job description for a science technician post.
   1.3 Describes the methods available for obtaining applicants for vacancies:
      (a) advertising;
      (b) direct recruitment from educational establishments;
      (c) personal contact.
   1.4 Discusses the designing of application forms.
   1.5 Describes methods of drawing up a short-list for interview.
   1.6 Discusses appropriate methods for interviewing candidates.
   1.7 Describes the analysis of the information obtained in 1.4—1.6 in the selection of a suitable candidate.
   1.8 Drafts a contract of employment.
   1.9 Drafts a letter of appointment.
   1.10 Describes procedures for terminating employment either at the wish of the employer or employee.

B EDUCATION AND TRAINING

2 Knows the principles and practices used in the education and training of laboratory staff.
   2.1 Discusses the differences between education and training.
   2.2 Defines the difference between in-post and external training.
   2.3 Describes the various educational and training courses and facilities available for science laboratory staff.
   2.4 Advises on courses relevant to the needs of the employer and employees in science laboratory work including:
      (a) courses leading to recognised qualifications;
      (b) short courses on specific topics organised in educational establishments;
      (c) specialised training courses organised by commercial/industrial enterprises;
      (d) conversion courses from one discipline to another.
   2.5 Discusses methods of providing effective in-post training.
   2.6 States the importance of retraining of experienced technical staff as the scientific discipline develops.
   2.7 Discusses the relationship between qualifications and staff grading.

C STAFF STRUCTURE

3 Understands the responsibilities of the supervisory technician in the area of staff relationships.
   3.1 Discusses the significance of grades in the staff structure and their inter-relationship.
   3.2 Explains the importance of career development.
   3.3 Discusses the basis on which recommendations for promotion may be made.
   3.4 Explains the need for discipline in the science laboratory environment.
   3.5 Discusses the importance of leadership and good example as the basis for disciplinary practice.

3.6 Explains the importance of encouragement in the maintenance of discipline.
3.7 Evaluates methods of criticism and of issuing reprimands.

D DECISION-MAKING

4 Understands the factors involved in the decision-making process.
   4.1 Describes processes by which problems are identified.
   4.2 Describes methods by which relevant information may be collated.
   4.3 Discusses techniques used to assess the information and to reach a decision.
   4.4 Explains the necessity for the decision-maker to take responsibility for his actions.
   4.5 Describes the structure of formal committees.
   4.6 Explains the necessity for written terms of reference for such committees.
   4.7 Explains the necessity for written standing orders.
   4.8 Describes the chain of decision-making.
   4.9 States the importance of corporate responsibility.
   4.10 Explains the need for committees to report to the appropriate authority.

E STAFF DEVELOPMENT

5 Understands the role of the supervisory technician in staff deployment.
   5.1 Estimates the establishment required for the efficient running of a laboratory unit.
   5.2 Explains the necessity for appropriate allocations of work-load and for job rotation.
   5.3 Explains the necessity for the allocation of specific tasks to be compatible with the employee's qualifications and experience.
   5.4 Describes the methods available for monitoring staff efficiency.

5.5 Explains the need for efficient communication and describes methods of achieving this.
5.6 Discusses the responsibilities placed upon line management at its various levels and the need for delegation of responsibility.
5.7 Explains the need for job satisfaction.

Acknowledgement is due to the Technician Education Council for permission to reproduce material from the standard units U78/512 and U78/513. The council reserves the right to revise the content of the units at any time. Further information and full details of the programmes may be obtained from the Technician Education Council at 76 Portland Place, London, W1N 4AA

Appendix 4
List of Papers Available from the Laboratories Investigation Unit

The Laboratories Investigation Unit is a group of architects, designers and others sponsored by the Department of Education and Science and the United Kingdom University Grants Committee, who have made a special study of laboratory design and building conversion.

The considerations and results of much of the unit's work have been published in the *Architects' Journal* and elsewhere and are likely to be of considerable interest to laboratory managers and others who are concerned with laboratory design and services.

Among the publications and reprints available are:

(1) An Approach to Laboratory Building (1969).
(2) Deep or Shallow Building: a comparison of costs in use (1970).
(3) Growth and Change in Laboratory Activity: a paper based on a study of the University of Surrey in Battersea (1971).
(4) The Economics of Adaptability: a comparative study of the initial and life costs of partitions (1971).
(5) Conversion of Buildings for Science and Technology. Part I (1971).
(6) Adaptable Furniture and Services for Education and Science (1972).
(7) Adaptable Laboratories — Practical Observations on Design and Installation (1974).

(8) Conversion of Buildings for Science and Technology. Part II (1977).
(9) Charles Darwin Building: Bristol Polytechnic (1977).
(10) Research Laboratories: Design for Flexibility (1977).

All of these papers (except No. 9 which costs £4) are available free from The Department of Education and Science, Architects and Building Branch, York Road, London, SE1 7PH.

Appendix 5
Safety Legislation in EEC Member States

Most European countries within the EEC have safety legislation similar to that in Britain, and safety considerations are included in company decisions. Employee participation in safety matters is given prominence and safety committees are an integral part of running businesses. Recent legislation includes:

Belgium. Law on Industrial Accidents, amended 1976. Requires notification of accidents at work to the Labour Inspector within 5 working days.

Denmark. The Working Environment Act 1975. Relates to administrative regulations drawn up in collaboration with the 12 safety committees set up for individual industries. Labour and management are both represented. Regulations also cover work done at home and with certification and marking of certain products and equipment. The Labour Inspection Service has issued safety instructions in connection with a number of industrial processes. Plans are in hand for training safety personnel.

France. The Council for the Prevention of Occupational Hazards is being set up with advisory powers in accident prevention and will consist of employees and technical specialists. Decrees are also issued on the medical supervision of workers exposed to lead, asbestos and other specific risks and on the co-ordination of health and safety measures.

Germany. Arbeitsstättenverordnung (Order Concerning Work Places) 1975. Mainly concerned with welfare of staff and provision of facilities such as washrooms. Another section relates to hazardous materials.

Ireland. EEC regulations on the classification, packaging and labelling of dangerous substances are being implemented and the existing Factories Act of 1955 is being improved in the area of health and welfare. Safety regulations are being updated to keep pace with technological advances in industry and include legislation covering eye protection and hazardous chemicals.

Italy. A national Health Committee is being established to replace the various independent bodies which exist at present. A parliamentary committee has been set up to investigate the recent dioxine disaster and to prepare regulations safeguarding the health of employees and the public.

Luxembourg. There is at present little or no legislation relating specifically to industrial safety but there are plans to set up occupational medical centres.

Netherlands. The Labour Act 1919, amended in 1977, is intended to protect young persons against health hazards. The Safety Decree for Factories and Workplaces has been updated to cope with technical advances in industry and covers aspects such as lighting, dangerous substances and noise. The Asbestos Decree 1977 prohibits the use and storage of crocidolite and imposes maximum limits on the concentration of asbestos fibres in the atmosphere of the workplace. Draft regulations are in existence concerned with workers' safety and welfare and with the organization of safety at work.

Index

Absenteeism, 201
Accident records, 151
Accidents, notifiable, 152
Accounts, 64
Acids, storage of, 84
Acknowledgement of order, 69
Activity charts, 196
Activity sampling, 196
Advertisements, 35
Agencies, advertising, 35
Agenda, 108
Air conditioning, 15
Air supply, filtered, 16
Alarms
    fire, 154
    temperature, 124
Alcohol
    duty on, 71
    purchase of, 71
    storage of, 71, 86
Allergy, 168
Alterations, building, 184
Ammonia, liquid, 87
Animal house safety, 131
Animal houses, 126
Animal returns, 130
Animal technicians, 130
Animals, 126
    cruelty to, 126, 129
    escape of, 132
    purchase of, 72
Apparatus
    servicing of, 178
    storage of, 88
Application forms, 36
Appointments, 45
Architects, 1, 184
Archives, 98
Audio-visual aids, 139
Autoclaves, 114
Automation, 186

Behaviour, human, 199
Benches, 12, 134

Biohazards, 172
Block release, 50
Booster pumps, water, 25
Brief, architect's, 1
Budgets, 62, 203
Building design, 1
Building maintenance, 184
Bulk buying, 78
Buying, 69

Cabinets, safety, 174
Calculators, 191
Capital equipment, 66
Carbon dioxide
    hazards of, 164
    storage of, 88
Carcinogens, 170
Card index, 99
Centrifuges, hazards of, 162, 179
Certificates, animal, 129
Chemicals
    hazards of, 166
    storage of, 82
Cine films, 141
Cine projectors, 141
Circulation lists, 206
Cleaning, premises, 181
Closed circuit television, 142
Co-carcinogens, 171
Codes of practice, 150
Cold-rooms, 123
Cold storage, 124
Columns, deionizing, 27
Committees, 107
Communicating, 202
Communications systems, 205
Computerization, 91, 95, 189
Computers, 189
Confirmation orders, 69
Consumables, purchase of, 67
Contracts of employment, 34, 45
Contracts, service, 178
Controlling, 202
Co-ordinating, 201

237

Index

Copying, document, 132
Copyright, 124, 134
Corrosives
  hazards of, 167
  storage of, 84
Costing, stores, 92
Credit notes, 70
Critical path analysis, 196
Cross-referencing, 99
Cruelty to Animals Act, 126, 129
Cryogenics, 87, 163
Cupboards, laboratory, 14
Cylinders, gas
  hazards of, 161
  storage of, 87

Dark rooms, 120
Data handling, 189
Data storage, 95, 187
Day release, 50, 131
Daylight utilization, 5
Decision-making, 104
Decontamination, 118
Decoration, 29, 183
De-ionized water, 26
Deliveries, 67
Delivery notes, 69
Demountable partitions, 5
Depreciation, 93
Dermatitis, 168
Design, laboratory, 1
Desks, 101
Detergents, 114, 168
Dewey decimal system, 100
Discipline, 56, 165
Discounts, 67, 73
Discrimination in employment, 45
Disinfection, 114
Dismissal, 58
Disposable items, 89, 115
Disposal, chemicals, 172
Distilled water, 26, 114
Document copying, 132
Documentation, stores, 90
Doors
  fire, 29, 158
  laboratory, 11
Dose, radiation, 118
Double glazing, 11
Drainage, 25
Drawer units, 14
Drying ovens, 110
Ducts
  service, 9
  ventilation, 16

Duplicators, 133
Duty
  alcohol, 71
  import, 66

EEC legislation, 234
Electric shock, 160
Electrical equipment, hazards of, 160
Equipment
  replacement of, 181
  selection of, 65
  servicing of, 178
Estimates, 62
Expenditure, 203

Fans, extract, 19
Filing, 97
Film badges, 118
Films, cine, 141
Filtered air supply, 16
Finance, sources of, 62
Finishes, 30
Fire, 152
Fire alarms, 154
Fire detectors, 154
Fire doors, 29, 158
Fire drill, 157
Fire escapes, 158
Fire extinguishers, 155
Fire-fighting equipment, 154
Fire prevention, 153, 159
First aid, 166
Flammable solvents, storage of, 85, 153
Floor coverings, 10, 114
Floors, 10
Flow diagrams, 196
Fluorescent lighting, 21
Forecasting, 193
Fume cupboards, 16, 117
Fume extract, 18
Furniture, 12
Further education, 48

Gas cylinders
  hazards of, 161
  rental of, 87
  storage of, 86
Gas supplies, 24, 135
Gases
  liquefied, 87, 164
  piped, 28
Genetic manipulation, 172

Index

Glassblowing, 143
Glassware sterilizing, 114
Glassware, storage of, 88
Glassware washing, 112
Grants, financial, 63

Health & Safety at Work Act, 146
Heat loss, 16
Heat sterilizing, 115
Heating, 15
High-speed centrifuges, 162
High-vacuum equipment, 166
High-voltage equipment, 160
Hot-rooms, 125
Hygiene, occupational, 167

Illumination, 21, 180
Import duty, 66, 72
Importation, 72
Income, sources of, 62
Indexing, 98
Induction, 46
Infection, prevention of, 173
Information
    distribution of, 203
    technical, 95, 208
Injuries, physical, 165
Inspection of equipment, 180
Inspection of premises, 179
Insurance, 93
Interviewing, 39
Invoices, 70
Ionizing radiations, 29
Irradiation, 115

Job description, 33
Job satisfaction, 199
Journals, 208

Laminates, plastic, 13, 30
Leadership, 56, 201
Leak testing, 87
Libraries, 97, 208
Licences, animal, 129
Lifting, 165
Lifts, 10, 127
Lighting, 21, 180
    emergency, 23
    workshop, 135
Liquefied gases, 87, 164
Load, electrical, 23
Low voltage supplies, 24

Maintenance, planned, 177
Mechanical services, 14
Medical history, 37
Meetings, 106, 207
Memoranda, 206
Method study, 196
Methylated spirit, purchase of, 71
Microbiological hazards, 172
Microfilm, 96
Minimum orders, 68
Minutes, 108
Mixer taps, 25
Modular design, 2
Monitoring, radiation, 118
Motivation, 53, 199
Multi-storey buildings, 30

Nitrogen, liquid, 87, 165
Noise, 134
Notice boards, 207

Occupational hygiene, 167
Office facilities, 101
Open days, 208
Operational research, 196
Order number, 69
Ordering, 68
Organizing, 195
Ovens
    drying, 112
    sterilizing, 114

Paint, 30
Partitions, demountable, 5
Pathogens, 173
Personnel selection, 39
Photographic materials, 120
Photographic units, 119
Physical injuries, 165
Pilfering, 74
Piped gases, 28
Planning, 194
Plastic laminates, 13, 30
Poisons
    purchase of, 71
    storage of, 85
Progress reports, 64
Project grants, 63
Projectors
    cine, 141
    overhead, 141
    slide, 140
Purchase orders, 69

Index

Purchasing, 65
Purchasing documents, 69

Quarantine, 127
Quinquennium, 63

Rabies Act, 127
Radiations, ionizing, 29, 161
Radioactive substances
    disposal of, 116
    purchase of, 73
    storage of, 88
Radioisotope laboratories, 115
Recorders, tape, 140
Record-keeping, 100, 117
Records, 193
Recruitment, 35
References, 37
Reprimands, 57
Reprographics, 132
Requisitions, 90
Resins, deionizing, 28
Ring mains, 23

Safety cabinets, 174
Safety committees, 149
Safety in design, 29
Safety in workshops, 138
Safety legislation, 146
Safety officers, 148
Sandwich courses, 51
Scrap metal, 139, 166
Screening, radiation, 29, 117
Screens, projection, 142
Selection boards, 40
Seminars, 207
Service charges, 178, 187
Service engineers, 178
Services, provision of, 9
Shortlisting, 42
Silver recovery, 120
Simian Herpes virus, 131
Sinks, 26
Skin reactions, 168
Slides, projection, 140
Solvents, flammable, storage of, 85
Space requirements, 4
Staff, selection of, 33, 39, 42
Staff turnover, 60, 200
Statement of account, 70
Steam supply, 28
Stencil copying, 133
Sterilization, 114

Stills, 27
Stock cards, 91
Stock lists, 93
Stocktaking, 93
Storage space, 4
Storekeeper, functions of, 79
Stores
    central, 77
    departmental, 77
    design of, 79
    management of, 76
    secondary, 77
    workshop, 135
Subcommittees, 109

Tape recorders, 140
TEC, 55, 223
Telephone orders, 69
Television, closed circuit, 142
Telex, 206
Temperature, air, 15
Termination of employment, 59
Terms of reference, 107
Think boxes, 103
Three-phase supply, 23, 135
Threshold limit values, 169
Time-lapse cine films, 141
Toxic materials, 167, 169
Training, 48, 131
Traps, waste, 26

Ultrasonic cleaning, 112
Ultraviolet irradiation, 115
Unfair dismissal, 58

Vacancies, staff, 36
Vaccination, 132
Vacuum equipment, 166
Valves
    gas, 29
    steam, 28
Velocity, air, 16
Ventilation, 15, 16, 117, 122, 128, 135
Vibration, 10, 134
Video tape, 142

Washing machines, glassware, 112
Waste pipes, 26
Water
    de-ionized, 26
    distilled, 26
    supplies, 25, 114

Windows, 11, 21
Work study, 195
Working groups, 200
Workshops, 134, 179

X-rays, 161

Zoonoses, 131